Food Microbiology and Food Safety

Series Editor:
Michael P. Doyle

More information about this series at http://www.springer.com/series/7131

Food Microbiology and Food Safety Series

The Food Microbiology and Food Safety series is published in conjunction with the International Association for Food Protection, a non-profit association for food safety professionals. Dedicated to the life-long educational needs of its Members, IAFP provides an information network through its two scientific journals (Food Protection Trends and Journal of Food Protection), its educational Annual Meeting, international meetings and symposia, and interaction between food safety professionals.

Xiangyu Deng • Henk C. den Bakker
Rene S. Hendriksen
Editors

Applied Genomics of Foodborne Pathogens

Springer

Editors
Xiangyu Deng
Center for Food Safety
University of Georgia
Griffin, GA, USA

Henk C. den Bakker
Department of Animal and Food Sciences
Texas Tech University
Lubbock, TX, USA

Rene S. Hendriksen
National Food Institute
Technical University of Denmark
Copenhagen, Denmark

Food Microbiology and Food Safety
ISBN 978-3-319-43749-1 ISBN 978-3-319-43751-4 (eBook)
DOI 10.1007/978-3-319-43751-4

Library of Congress Control Number: 2016954648

Printed on acid-free paper

This Springer imprint is published by Springer Nature
The registered company is Springer International Publishing AG
The registered company address is: Gewerbestrasse 11, 6330 Cham, Switzerland

Preface

While the first two complete bacterial genome sequences were published in 1995 and whole genome sequencing (WGS) has since changed the landscape of microbiology, this technique had long been a privilege of sequencing centers and mainly served the purpose of scientific research. The applied use of WGS, for example, in the field of food safety and public health, had been prohibitive due to its high cost, lengthy procedure, and technical challenges in data analysis.

Recent advances in so-called next-generation sequencing (NGS) technologies have transformed the once centralized resource into a viable and practical tool for microbial identification and characterization, which is becoming increasingly accessible to individual laboratories around the world. The democratization of WGS has opened new avenues for studying, tracking, and controlling foodborne pathogens, which, as reflected in this book, is most evident in public health surveillance and outbreak investigation of foodborne infectious diseases.

By providing a timely summary of recent proceedings, case studies, opinions, and trends, we hope that this book will present to food safety and public health professionals a snapshot of the emerging and fast-developing field of applied genomics of foodborne pathogens. It is of course impossible to exhaustively summarize all aspects of this field. And like any book on the topics of genomics and bioinformatics, it is challenging to keep up with the latest developments due to the fast-evolving nature of the technologies and disciplines. In this sense, this book would best serve as a guide and a stepping stone to a large and increasing body of literature, many of which are works from contributors of this book.

We are deeply grateful to our contributors who are frontline practitioners, leading subject matter experts, and critical players in the exciting endeavor of transforming public health microbiology with WGS.

We greatly appreciate the vision, support, cooperation, and tolerance of the staff of Springer Nature—in particular our executive editor Susan Safren and our project coordinator Michael Koy, who managed the entire project and made this book possible.

Griffin, GA, USA — Xiangyu Deng
Lubbock, TX, USA — Henk C. den Bakker
Copenhagen, Denmark — Rene S. Hendriksen

Contents

Contributors

Frank M. Aarestrup National Food Institute, Technical University of Denmark, Lyngby, Denmark

Department of Microbiology and Infection Control, Statens Serum Institut, Copenhagen, Denmark

Johanne Ahrenfeldt Department of Systems Biology, Center for Biological Sequence Analysis, Technical University of Denmark, Lyngby, Denmark

Henk C. den Bakker Department of Animal and Food Sciences, Texas Tech University, Lubbock, TX, USA

David Boxrud Minnesota Department of Health, St. Paul, MN, USA

Alfredo Caprioli European Union Reference Laboratory for Escherichia coli, Department of Veterinary Public Health and Food Safety, Istituto Superiore di Sanità, Rome, Italy

Heather Carleton Enteric Diseases Laboratory Branch, Centers for Disease Control and Prevention, Atlanta, GA, USA

Jose Luis Bellod Cisneros Department of Systems Biology, Center for Biological Sequence Analysis, Technical University of Denmark, Lyngby, Denmark

Salvatore Cosentino Department of Infection Metagenomics, Genome Information Research Center, Research Institute for Microbial Diseases, Osaka University, Suita, Osaka, Japan

Xiangyu Deng Center for Food Safety, University of Georgia, Griffin, GA, USA

Peter Gerner-Smidt Enteric Diseases Laboratory Branch, Centers for Disease Control and Prevention, Atlanta, GA, USA

Henrik Hasman Department of Microbiology and Infection Control, Statens Serum Institut, Copenhagen, Denmark

Rene S. Hendriksen National Food Institute, Technical University of Denmark, Lyngby, Denmark

Katrine G. Joensen Department of Microbiology and Infection Control, Statens Serum Institut, Copenhagen, Denmark

Vanessa Jurtz Department of Systems Biology, Center for Biological Sequence Analysis, Technical University of Denmark, Lyngby, Denmark

Rolf Sommer Kaas National Food Institute, Technical University of Denmark, Lyngby, Denmark

Marion P.G. Koopmans Department of Viroscience, Erasmus Medical Centre, Rotterdam, The Netherlands

Virology Division, Centre for Infectious Diseases Research, Diagnostics and Screening, National Institute for Public Health and the Environment, Bilthoven, The Netherlands

Mette Voldby Larsen Department of Systems Biology, Center for Biological Sequence Analysis, Technical University of Denmark, Lyngby, Denmark

Pimlapas Leekitcharoenphon National Food Institute, Technical University of Denmark, Lyngby, Denmark

Oksana Lukjancenko Department of Systems Biology, Center for Biological Sequence Analysis, Technical University of Denmark, Lyngby, Denmark

National Food Institute, Technical University of Denmark, Lyngby, Denmark

Ole Lund National Food Institute, Technical University of Denmark, Lyngby, Denmark

Department of Systems Biology, Center for Biological Sequence Analysis, Technical University of Denmark, Lyngby, Denmark

Noel McCarthy Warwick Medical School, University of Warwick, Coventry, UK

Valeria Michelacci European Union Reference Laboratory for Escherichia coli, Department of Veterinary Public Health and Food Safety, Istituto Superiore di Sanità, Rome, Italy

Stefano Morabito European Union Reference Laboratory for Escherichia coli, Department of Veterinary Public Health and Food Safety, Istituto Superiore di Sanità, Rome, Italy

Justin O'Grady Norwich Medical School, University of East Anglia, Norwich, UK

Thomas Nordahl Petersen Department of Systems Biology, Center for Biological Sequence Analysis, Technical University of Denmark, Lyngby, Denmark

Simon Rasmussen Department of Systems Biology, Center for Biological Sequence Analysis, Technical University of Denmark, Lyngby, Denmark

Louise Roer National Food Institute, Technical University of Denmark, Lyngby, Denmark

Joelle K. Salazar Division of Food Processing Science and Technology, U. S. Food and Drug Administration, Bedford Park, IL, USA

Dhany Saputra Department of Systems Biology, Center for Biological Sequence Analysis, Technical University of Denmark, Lyngby, Denmark

Jørgen Schlundt Nanyang Technological University, Singapore, Singapore

Thomas Sicheritz-Ponten Department of Systems Biology, Center for Biological Sequence Analysis, Technical University of Denmark, Lyngby, Denmark

Saskia L. Smits Department of Viroscience, Erasmus Medical Centre, Rotterdam, The Netherlands

Laura K. Strawn Department of Food and Technology, Eastern Shore Agricultural Research and Extension Center, Painter, VA, USA

Martin Christen Frølund Thomsen Department of Systems Biology, Center for Biological Sequence Analysis, Technical University of Denmark, Lyngby, Denmark

Rosangela Tozzoli European Union Reference Laboratory for Escherichia coli, Department of Veterinary Public Health and Food Safety, Istituto Superiore di Sanità, Rome, Italy

Eija Trees Enteric Diseases Laboratory Branch, Centers for Disease Control and Prevention, Atlanta, GA, USA

John Wain Norwich Medical School, University of East Anglia, Norwich, UK

Yun Wang Division of Food Processing Science and Technology, U. S. Food and Drug Administration, Bedford Park, IL, USA

Peter R. Wielinga National Food Institute, Technical University of Denmark, Lyngby, Denmark

William J. Wolfgang Wadsworth Center/New York State Department of Health, Albany, NY, USA

Ea Zankari Department of Systems Biology, Center for Biological Sequence Analysis, Technical University of Denmark, Lyngby, Denmark

National Food Institute, Technical University of Denmark, Lyngby, Denmark

Wei Zhang Department of Food Science and Nutrition, Illinois Institute of Technology, Bedford Park, IL, USA

Chapter 1
Role of Whole Genome Sequencing in the Public Health Surveillance of Foodborne Pathogens

Peter Gerner-Smidt, Heather Carleton, and Eija Trees

Introduction

Public health departments have a critical role in promoting health and preventing illness and injury in the society. They do this through ongoing surveillance and investigation of public health emergencies such as outbreaks. This way, risk and protective factors for illnesses and injuries are identified and recommendations to avoid and prevent them may be made to regulators and other decision makers.

Public health laboratories are crucial for the preparedness and response to bacterial foodborne infections by providing data for laboratory based surveillance, laboratory confirmatory data for other surveillance systems, e.g., clinical notification, and confirmation of the identity of isolates submitted from clinical laboratories, and detection and isolation of foodborne pathogens from clinical specimens, and food samples of public health importance not handled by other laboratories.

Current Workflows in the Public Health Laboratories

Clinical laboratories typically submit specimens and isolates to the public health laboratories as part of voluntary or legally mandated laboratory surveillance but also for reference testing to assist them in identifying or confirming the identity of pathogens they have isolated. The public health laboratories typically identify the genus

P. Gerner-Smidt (✉) • H. Carleton • E. Trees
Enteric Diseases Laboratory Branch, Centers for Disease Control and Prevention, Atlanta, GA, USA
e-mail: plg5@cdc.ogv; wvt2@cdc.gov; eih9@cdc.gov

X. Deng et al. (eds.), *Applied Genomics of Foodborne Pathogens*, Food Microbiology and Food Safety, DOI 10.1007/978-3-319-43751-4_1

and species of all the isolates they receive. Additionally, the isolates are further characterized in different ways as part of the laboratory surveillance specific to the pathogens in question, e.g., serotyping of *Salmonella*, Shiga toxin-producing *E. coli* (STEC), *Shigella*, *Vibrio* and *Listeria*; virulence characterization of e.g., any diarrheagenic *E. coli* pathotype (STEC, enteropathogenic *E. coli* [EPEC], enterotoxigenic *E. coli* [ETEC], enteroaggregative *E. coli* [EAEC], enteroinvasive *E. coli* [EIEC] and *Shigella*) or *Vibrio* spp.; antimicrobial susceptibility testing of *Salmonella*, *E. coli*, *Campylobacter*, and other pathogens; high discriminatory subtyping for outbreak detection and investigation. Many different methods of varying complexity are used for these purposes. They include among others growth characteristics on simple and complex media, fermentation tests and other biochemical reactions, agglutination with diagnostic antisera, immunofluorescence and other immune assays, cell culture assays, protein electrophoretic assays and molecular assays. As the genetic determinants for the different phenotypic tests have been identified, some of them have been replaced by molecular PCR or DNA hybridization assays directed at these targets; such tests have because of their simplicity and accuracy since the 1980s replaced traditional methods for species identification, serotyping and virulence characterization of foodborne pathogens in many public health laboratories. Such assays are also increasingly used for screening for multiple pathogens in the clinical specimens, e.g., in stool samples in order to identify specimens most likely to contain pathogens for culture. These diagnostic panels have also been introduced commercially in the clinical laboratories for culture independent diagnostics of enteric infections in recent years.

In the most recent decades molecular methods have taken over for routine high discriminatory subtyping of foodborne pathogens for outbreak detection and investigation; pulsed field gel-electrophoresis (PFGE) is a nearly universal method that has become the de facto standard for outbreak surveillance of almost any bacterial pathogen and has been used in PulseNet [1], the national subtyping network for molecular surveillance of foodborne infections in the United States, since 1996. PFGE has in recent years been supplemented with multi locus variable number of tandem repeats analysis (MLVA) for specific pathogens [2], e.g., *E. coli* O157, *Salmonella* ser. Typhimurium and Enteritidis.

Since foodborne infections do not respect any borders because of international trade and travel, an infection detected in one country may have its origin in a country thousands of miles away, and a broadly distributed food may cause outbreaks in multiple countries at the same time. In order to detect and investigate international outbreaks it is critically important that public health laboratories collaborate and use the same methods for characterization and subtyping of foodborne pathogens [3]. Most serotyping schemes, e.g., the Kaufmann–White scheme for *Salmonella*, are therefore standardized internationally and most public health laboratories doing high discriminatory subtyping of foodborne pathogens use the PulseNet protocols for PFGE or protocols that are compatible with them.

Most methods have been developed and are used for the characterization of a single or a few pathogens, only. The turn-around time for the full characterization in the public health laboratories ranges typically from 4 days to several months depending on the pathogen and the strain. Thus, the workflows in the public health laboratories are many, variable and complex and require substantial subject matter expertise about each pathogen under surveillance in order to be able to interpret the results correctly. Laboratory surveillance is for these reasons today extremely complex and costly. Most of the new methods that have been introduced in the public health in the past have added new workflows to the surveillance and have therefore not resulted in cost savings or simplification of workflows.

The Ideal Public Health Laboratory System

In the public health laboratory ideally all work with every pathogen should be conducted in a single, simple workflow. All information generated should be backwards compatible with older methods in order not to lose any historical information. The information should include data of all degrees of complexity and differentiation capabilities from simple identification of the genus and species at the population level to specific strain level characteristics used for outbreak detection and investigation. Data generated in one laboratory must be compatible with data on the same strains generated in any other laboratory to ensure efficient detection of local, national or global outbreaks. For the same reasons, data should be shared freely among the public health laboratories. In order to ensure efficient communication about the spread of specific clones of foodborne pathogens and for outbreak investigations, the nomenclature of strain types must be stable, standardized and internationally recognized. The methods must be user-friendly, fast, economical and equally applicable to local, national and international investigations. Since public health microbiologists in general have basic laboratory skills and often little understanding of bioinformatics and high performance computing (HPC) is not available in most public health laboratories, analytical tools for data must be tailored to this situation, i.e. the analysis should be designed to be used by people with limited insight into complex analytical approaches and without the need for high performance computing. For the most basic needs a black box approach is sufficient but the data generated should also be available for more detailed mining by analytical experts, e.g., bioinformaticians. Finally, data generated at all levels should have a high level of epidemiological concordance.

Only a few of these requirements are fulfilled in public health laboratories today. However, whole genome sequencing (WGS) shows promise as a method for strain characterization that could fulfill all or almost all of these criteria.

Whole Genome Sequencing in Public Health

Many of the WGS analytical tools for reference identification and characterization of foodborne pathogens have already been developed and are available to the scientific community on the web, e.g., those available in the Center for Genomic Epidemiology (CGE) tool box (https://cge.cbs.dtu.dk/services/) described in a different chapter in this book [4]: SpeciesFinder, a 16s rRNA-based tool for species identification; VirulenceFinder, that identifies virulence genes and their variants in *E. coli*; PlasmidFinder, a tool for the identification of plasmids in *Enterobacteriaceae*; ResFinder that identifies acquired resistance determinants to name a few. Similar tools have also been developed for molecular serotyping of *Salmonella* serotypes [5] and *E. coli* serotypes (SeroType Finder; https://cge.cbs.dtu.dk/services/SerotypeFinder/). A drawback of any molecular system developed to predict a specific phenotype is that not all genes are expressed or more than one gene representing different phenotypes, e.g., serotypes, might be present in a strain but only one of them is expressed confusing the interpretation of the result [5]. For the most part, the issue of expression of the genes detected is not a problem but for surveillance it may be critical to know at least the proportion of genes that might not be expressed. For example, for antimicrobial resistance monitoring it will be critical to establish a sentinel surveillance system, where a random representative sample of the isolates are tested by traditional susceptibility testing methods in addition to WGS to detect changes in expression that leads changes in therapeutic options and to detect hitherto unknown resistance mechanisms since the genotypic system can only detect resistance markers that are already known.

Many schemes for multi locus sequence typing (MLST) based on a few housekeeping or virulence genes were developed in the 1990s and 2000s [6]. Originally, the alleles were determined by sequencing short PCR amplicons of the genes/loci used in these schemes. Today it is cheaper to sequence the whole genome and extract the gene sequences than to amplify and sequence the specific sequences for each scheme. For this reason, MLST today is performed by WGS in most laboratories.

Many of the web-sites allow batching of isolates to be characterized. However, in order to characterize an isolate by more tools, the user will have to log in to each tool separately. This limits the utility of the tools in the public health routine workflow where isolates typically are queried for multiple characteristics since it is cumbersome to query multiple sites separately.

So far, WGS has mostly been used for subtyping to investigate outbreaks in public health laboratories but not to detect them. Exceptions are recently established surveillance of listeriosis in Denmark (http://www.ssi.dk/English/News/EPI-NEWS/2014/No%2018%20-%202014.aspx) and the United States (http://www.cdc.gov/amd/project-summaries/listeria.html) and a pilot study on surveillance of STEC in Denmark [7]. Most experience comes from retrospective studies which may indicate the potential of the use of WGS for outbreak investigations but not prove the actual public health impact of the new technology [8–13]. Another weakness of the retrospective investigations is that the data can only be interpreted with existing epidemiological information available whereas when used in a real-time the WGS may be used to guide the epidemiological investigation thereby further illustrating the power of the tool. However, all studies so far have shown that WGS has equal or better resolution than the current reference methods, e.g., PFGE and MLVA. This is also true for very clonal pathogens, e.g., *Salmonella enterica* ser. Enteritidis [14, 15]. Theoretically, restriction polymorphisms as detected by PFGE or variations in the number of tandem repeats at different loci as detected by MLVA may be predicted from WGS; however, in practice this is not possible since accurate assemblies of the whole genome are needed to predict both the PFGE and MLVA profiles and for PFGE the genome sequence ideally needs to be closed, which rarely happens. Therefore, other approaches based on single nucleotide polymorphism (SNP) analysis or MLST analysis of all genes (whole genome [wg] MLST) or the core genes (core genome [cg] MLST) present in the isolates have been used.

Unlike PFGE and MLVA, SNP and MLST data are phylogenetically informative, i.e. represent evolution. This is helpful when assessing possible epidemiological relationships of isolates that show variations in their subtype, as phylogenetic relatedness cannot be inferred from PFGE patterns alone without epidemiological information that suggest a relationship.

So far, SNP-based approaches have mostly been used for outbreak investigations [8, 9, 11, 12]. Their strengths are that they are very powerful and can be used if one has at least one good reference sequence to align the SNP's against, unless a reference-free approach is performed like kSNP or kmer's, though kSNP and kmer's have lower resolution than reference based SNP methods. Open source software may be used as well as a number of commercial "off-the-shelf" software. The current weaknesses of the SNP-based approaches are that they require specific bioinformatics skills from the person performing the analysis, access to high performance computers, either in house or via broadband Internet; additionally, analysis can be slow to perform depending on the number of isolates compared Results generated using different analytical pipelines or with SNP's aligned against different reference sequences are not directly comparable and for that reason the SNP-profile of a strain cannot be named unambiguously, which makes communication of results with other actors in the outbreak investigation and the public difficult. Because of the current complex analysis, need of bioinformatics expertise and lack of a stable nomenclature, SNP analysis is suboptimal for routine use in public health. However, the approach may be used when other sequence-based approaches, e.g., MLST approaches, have not been developed or do not have adequate resolution.

The current alternative to the SNP-analysis is MLST analysis (Table 1.1).

MLST analysis in its most extensive forms, cgMLST and wgMLST, are extremely powerful subtyping tools with a discriminatory power that does not seem to be any

Table 1.1 Key features of SNP and MLST approaches of importance to public health

	SNP approaches	MLST approaches
Epidemiological concordance	High	High
Stable nomenclature	No	Yes
Reference characterization: Identification, serotyping, virulence and antimicrobial resistance markers	No	Yes
Speed	Slow SNP calling, slow analysis	Slow allele calling, fast analysis
Local computing requirements	Medium–high	Low
Local bioinformatics expertise	Yes	No
Curation of database	No	Yes

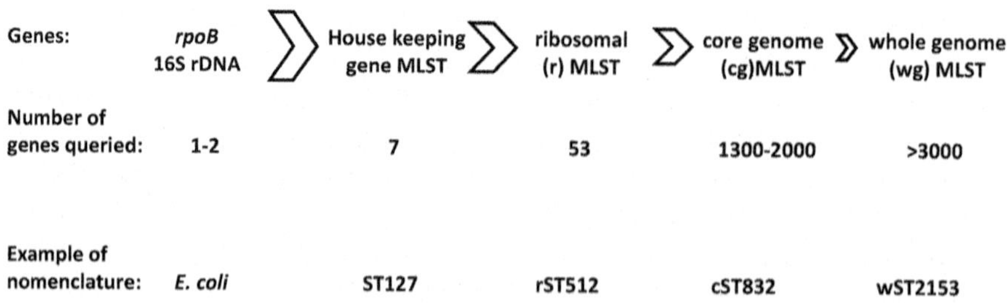

Fig. 1.1 Tiered characterization of foodborne pathogens by whole genome sequencing using a gene–gene/MLST approach

different from that of SNP analysis and hence suitable for high discriminatory subtyping [6, 13, 16, 17]. Since the method is assessing allelic variations in the genes, it is possible to tailor the discriminatory power of the system to fundamentally different surveillance activities, e.g. trend analysis, microbiological food source attribution or outbreak detection, by designing systems that include different numbers of genes/loci [6]; the discriminatory power increases as the number of genes/loci are increased, e.g., traditional house-keeping gene MLST that typically assess variations in seven genes has fairly low discriminatory power, ribosomal RNA MLST (rgMLST) that includes 53 genes has higher discrimination, and schemes including all the core genes or the full pan genome (cgMLST or wgMLST) have the highest discriminatory power (Fig. 1.1).

Since each subtype at every level is defined by a well characterized set of alleles, each subtype may be named unambiguously with a name that will not change as more strains are added to the database or comparisons change, i.e., MLST leads to stable nomenclature, which is highly desirable for communication purposes in public health. Additionally, by designing a system of subtypes by increasing discriminatory power, the relatedness of isolates may be depicted by comparing their names, i.e., the more sequence types shared between isolates, the closer related they are. Another advantage of this tiered MLST approach is that it provides more options for subtyping for different purposes, e.g., while cgMLST or wgMLST might be optimal for outbreak detection and investigations, less resolution using 7-gene MLST or rgMLST might be more appropriate for attribution analysis (predicting the food sources of sporadic infections from comparison of the subtypes of clinical isolates with those from food production isolates) [18]. Another advantage of the MLST approach is that once the alleles have been identified and stored in a database, analysis may be performed on standard low-capacity laboratory computers using commercial over the shelf software (COTS), e.g, SeqSphere+(Ridom® GmbH, Münster, Germany), or BioNumerics (Applied Maths, Austin, TX) or using web-applications e.g. PubMLST (http://pubmlst.org/databases.shtml) by personnel with little or no bioinformatics expertise. A current limitation of the MLST approach are that a comprehensive allele database of genes representing the full diversity of the organism needs to be constructed before it can be used and that this database needs to be curated to identify, confirm and name new alleles. Building the allele database is no trivial task since it typically is based on high quality, assembled sequences of often several 100 reference strains. Though, once the database has been created, alleles of test isolates may be assigned either from raw reads or assemblies and most curation of the databases may be automated. However, to be useful for international surveillance the database will need to be standardized internationally with standardization of quality measures for allele and locus definitions. Establishing such databases is under way in a collaboration between PulseNet International and partners in international reference laboratories and academia for *Listeria monocytogenes*, *Campylobacter*, *E. coli* and *Shigella*, and *Salmonella*. International collaboration is not just necessary because of the need for international standardization but also because the task is so big and resource intensive that no single institution may lift it alone.

Combining Reference Characterization and Subtyping in Whole Genome Sequencing Based System for Public Health

In the ideal system for public health laboratory surveillance, raw WGS data are automatically submitted to and stored in a public repository, e.g., Sequence Read Archive (SRA) at National Center for Biotechnology Information (NCBI). Next the raw reads are automatically trimmed, assembled as necessary and gene by gene information is extracted, named and returned to the public health user, who may further analyze the data for public health purposes, e.g., outbreak detection and investigation. A practical setup is shown in the diagram in Fig. 1.2.

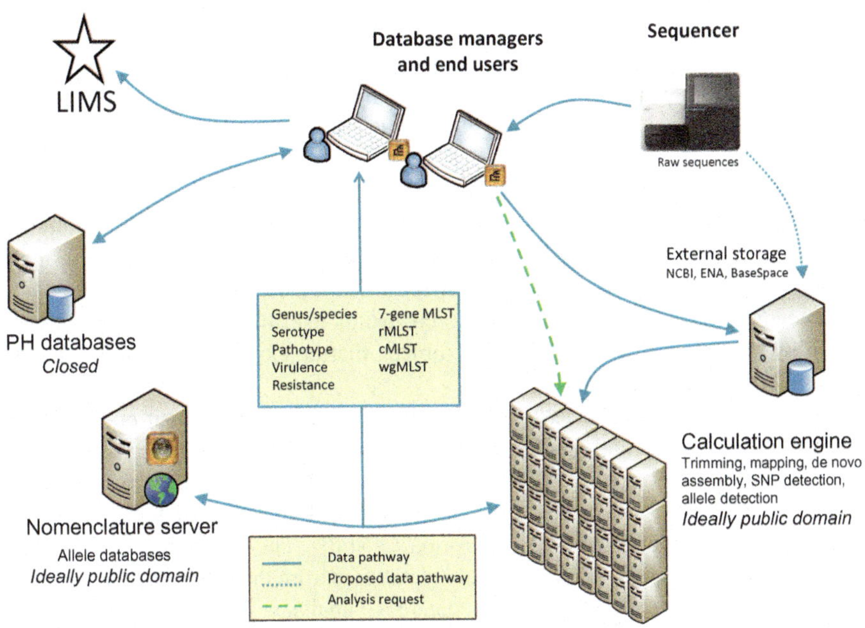

Fig. 1.2 Whole genome sequencing workflow in a public health laboratory network working with foodborne pathogens

When a sample is loaded to the sequencer, all metadata related to the sample and relevant to public health is entered into the public health database. Once the isolate has been sequenced and the output passes defined quality criteria, the raw sequences are uploaded to a public repository, e.g., SRA at NCBI, along with as many associated metadata as possible, excepting confidential information and information that might hamper the public health response if shared publically. Then, analysis requests are submitted by an end-user or automatically to a high performance computing cluster, the 'calculation engine', which downloads the raw reads from the public repository, then trims the reads, assembles them as necessary and the relevant gene and allele information is extracted. The extracted information is then passed through a curated database that contains all sequence information about known genes and alleles ('nomenclatural server'), and translates the sequence information to gene, gene variant (e.g., for genus/species identification, serotyping, virulence characterization and antimicrobial resistance marker identification) and allele names (for MLST analysis) which are then transferred to the local public health database, which contains all the associated metadata needed for public health action. Specific reference characteristics of the isolates may then also be transferred to a LIMS-system if these data are to be reported to the laboratory that submitted the isolate if the laboratory does not have direct access to the public health database. New alleles are named automatically in the nomenclatural server if specific quality criteria are met, or the sequences are flagged for manual identification and naming by the curators of the national database. The curators of the national databases also perform cluster searches to identify national clusters that would not be identified by local end-users. The end-user may also perform SNP analysis on the reads through the calculation engine if the MLST analysis does not provide unequivocal answers.

The calculation engine and the nomenclatural server should ideally be placed in public domain. Neither contains confidential information and free sharing of the data on the nomenclatural server is critical to ensure international compatibility and the use of the same unified nomenclature globally. These resources will this way be available to lesser resourced public health laboratories and to the whole scientific community. Ideally, the data in these resources should be in a format that does not require the use of specific software. The nomenclatural database needs to be curated in real-time to ensure its usefulness for outbreak investigations and the curators should be subject matter experts from international reference laboratories.

The public health databases are closed and only accessible for their public health users with one in each national/local public health laboratory and in regional public health institutions as needed. The data in the public health databases may be analyzed on standard laboratory computers using standard MLST and database software by the end-user with no specific bioinformatics skills. The raw sequence data may also be extracted through the public health databases from the repository for research purposes or for more specific analysis, e.g., SNP analysis, when the MLST fails or produce ambiguous results.

Conclusions

The system described above fulfills all requirements for the ideal public health isolate characterization system except for backwards compatibility with PFGE and MLVA. However, as part of the construction and validation of the nomenclatural MLST database numerous well characterized strains from previous outbreaks and sporadic isolates representing the full diversity of foodborne pathogens will be analyzed and sequenced thereby enabling correlation of the most important PFGE and MLVA profiles to MLST clones. The system may with the few reservations regarding possible lack of expression of some genes, e.g., resistance genes, replace all the characterization of foodborne pathogen cultures in most public health laboratories. This will represent a paradigm shift in public health microbiology and lead to cost savings compared to the existing surveillance. It is estimated that WGS represents a cost savings of characterizing *Campylobacter* and Shiga toxin-producing *E. coli* (STEC) of 50–55 % considering reagent and supply costs alone (Table 1.2).

The savings might not be as big for all pathogens but the amount of information that will be made available for surveillance, all in real time, will dramatically increase, e.g., antimicrobial resistance and subtyping data will become available on all isolates. This will greatly improve the utility of the data: more outbreaks will be detected and with the higher resolution provided by WGS, isolates that could not be separated by PFGE or MLVA may be differentiated thus helping to focus limited epidemiological resources to clusters that are most likely to represent outbreaks; microbiological attribution of the most important food sources of sporadic infections will become possible; and it will become possible to more accurately follow trends of specific organisms, plasmids or transmissible traits, e.g., antimicrobial resistance or virulence genes enabling early recognition of emerging problems so they may be addressed faster; finally, our understanding of the epidemiology of foodborne illnesses will greatly increase through correlation of systematically collected epidemiological information with the sequence information. This way new microbiological

Table 1.2 Costs of supplies for routine characterization of *Campylobacter* and Shiga toxin-producing *E. coli* (STEC) using traditional methods and whole genome sequencing (WGS) using an MiSeq sequencer (Illumina Inc, San Diego, CA, USA)

Characterization	*Campylobacter* (traditional)	*Campylobacter* (WGS)	STEC (traditional)	STEC (WGS)
Identification	$74.20		$60	
Serotyping			$159	
Virulence PCR			$10	
MLST	$71.80			
PFGE			$30	
MLVA			$15	
Total	$146.00	$73	$274	$123
Cost savings		**50 %**		**55 %**

risk factors or potential targets for vaccines to control foodborne disease could be identified thereby facilitating their control. The future is here.

Disclaimers
The findings and conclusions in this chapter are those of the author and do not necessarily represent the official position of the Centers for Disease Control and Prevention.

Use of trade names is for identification only and does not imply endorsement by the Centers for Disease Control and Prevention or by the U.S. Department of Health and Human Services.

References

1. Swaminathan B, Barrett TJ, Hunter SB, Tauxe RV, The CDC PulseNet Task Force. PulseNet: the molecular subtyping network for foodborne bacterial disease surveillance, United States. Emerg Infect Dis. 2001;7:382–9.
2. Nadon CA, Trees E, Ng LK, Moller Nielsen E, Reimer A, Maxwell N, et al. Development and application of MLVA methods as a tool for inter-laboratory surveillance. Euro Surveill. 2013;18(35):20565.
3. Swaminathan B, Gerner-Smidt P, Ng LK, Lukinmaa S, Kam KM, Rolando S, et al. Building PulseNet International: an interconnected system of laboratory networks to facilitate timely public health recognition and response to foodborne disease outbreaks and emerging foodborne diseases. Foodborne Pathog Dis. 2006;3(1):36–50.
4. Larsen MV, Joensen KG, Zankari E, Ahrenfeldt J, Lukjancenko O, Kaas RK, et al. The CGE tool box. In: Deng X, den Bakker H, Hendriksen R, editors. Applied genomics of foodborne pathogens. New York, NY: Springer; 2015.
5. Zhang S, Yin Y, Jones MB, Zhang Z, Deatherage Kaiser BL, Dinsmore BA, et al. *Salmonella* serotype determination utilizing high-throughput genome sequencing data. J Clin Microbiol. 2015;53(5):1685–92.
6. Maiden MC, Jansen van Rensburg MJ, Bray JE, Earle SG, Ford SA, Jolley KA, et al. MLST revisited: the gene-by-gene approach to bacterial genomics. Nat Rev Microbiol. 2013;11(10):728–36.
7. Joensen KG, Scheutz F, Lund O, Hasman H, Kaas RS, Nielsen EM, et al. Real-time whole-genome sequencing for routine typing, surveillance, and outbreak detection of verotoxigenic *Escherichia coli*. J Clin Microbiol. 2014;52(5):1501–10.
8. Allard MW, Luo Y, Strain E, Li C, Keys CE, Son I, et al. High resolution clustering of *Salmonella enterica* serovar Montevideo strains using a next-generation sequencing approach. BMC Genomics. 2012;13(1):32.
9. Gilmour MW, Graham M, Van Domselaar G, Tyler S, Kent H, Trout-Yakel KM, et al. High-throughput genome sequencing of two *Listeria monocytogenes* clinical isolates during a large foodborne outbreak. BMC Genomics. 2010;11(1):120.
10. McDonnell J, Dallman T, Atkin S, Turbitt DA, Connor TR, Grant KA, et al. Retrospective analysis of whole genome sequencing compared to prospective typing data in further informing the epidemiological investigation of an outbreak of *Shigella sonnei* in the UK. Epidemiol Infect. 2013;21:1–8.
11. Mellmann A, Harmsen D, Cummings CA, Zentz EB, Leopold SR, Rico A, et al. Prospective genomic characterization of the German enterohemorrhagic *Escherichia coli* O104:H4 outbreak by rapid next generation sequencing technology. PLoS One. 2011;6(7), e22751.
12. Reimer AR, Van Domselaar G, Stroika S, Walker M, Kent H, Tarr C, et al. Comparative genomics of *Vibrio cholerae* from Haiti, Asia, and Africa. Emerg Infect Dis. 2011;17(11):2113–21.

13. Schmid D, Allerberger F, Huhulescu S, Pietzka A, Amar C, Kleta S, et al. Whole genome sequencing as a tool to investigate a cluster of seven cases of listeriosis in Austria and Germany, 2011–2013. Clin Microbiol Infect. 2014;20(5):431–6.
14. Allard MW, Luo Y, Strain E, Pettengill J, Timme R, Wang C, et al. On the evolutionary history, population genetics and diversity among isolates of *Salmonella* Enteritidis PFGE pattern JEGX01.0004. PLoS One. 2013;8(1), e55254.
15. Deng X, Shariat N, Driebe EM, Roe CC, Tolar B, Trees E, et al. Comparative analysis of subtyping methods against a whole-genome-sequencing standard for *Salmonella enterica* serotype Enteritidis. J Clin Microbiol. 2015;53(1):212–8.
16. Kohl TA, Diel R, Harmsen D, Rothganger J, Walter KM, Merker M, et al. Whole-genome-based *Mycobacterium tuberculosis* surveillance: a standardized, portable, and expandable approach. J Clin Microbiol. 2014;52(7):2479–86.
17. Leopold SR, Goering RV, Witten A, Harmsen D, Mellmann A. Bacterial whole-genome sequencing revisited: portable, scalable, and standardized analysis for typing and detection of virulence and antibiotic resistance genes. J Clin Microbiol. 2014;52(7):2365–70.
18. Ogden ID, Dallas JF, MacRae M, Rotariu O, Reay KW, Leitch M, et al. Campylobacter excreted into the environment by animal sources: prevalence, concentration shed, and host association. Foodborne Pathog Dis. 2009;6(10):1161–70.

Chapter 2
Global Microbial Identifier

Peter R. Wielinga, Rene S. Hendriksen, Frank M. Aarestrup, Ole Lund, Saskia L. Smits, Marion P.G. Koopmans, and Jørgen Schlundt

Introduction

Human and animal populations increasingly share a number of emerging and re-emerging infections including infections that are exchanged between these populations (i.e. zoonotic infections) either directly or indirectly through food or vectors. Recent global outbreaks, such as SARS (Severe Acute Respiratory Syndrome), avian influenza (H5N1), pandemic (swine)influenza (H1N1) and MERS (Middle East Respiratory Syndrome) have rightfully received global attention, both in relation to the disease burden, the risk of rapid spread and the additional economic cost relative to travel and trade restrictions. To complete the picture of the disease burden and economic cost of human disease related to animals a number of endemic human infections that are continuously transferred from animals (e.g. salmonellosis, brucellosis, campylobacteriosis, rabies, cysticercosis) should also be considered. It is estimated that more than six out of every ten emerging infectious diseases in humans are

P.R. Wielinga • R.S. Hendriksen • F.M. Aarestrup • O. Lund
National Food Institute, Technical University of Denmark, Lyngby, Denmark
e-mail: peter.wielinga@gmail.com; rshe@food.dtu.dk; fmaa@food.dtu.dk; lund@cbs.dtu.dk

S.L. Smits
Department of Viroscience, Erasmus Medical Centre, Rotterdam, The Netherlands
e-mail: s.smits@erasmusmc.nl

M.P.G. Koopmans
Department of Viroscience, Erasmus Medical Centre, Rotterdam, The Netherlands

Virology Division, Centre for Infectious Diseases Research, Diagnostics and Screening, National Institute for Public Health and the Environment, Bilthoven, The Netherlands
e-mail: m.koopmans@erasmusmc.nl

J. Schlundt (✉)
Nanyang Technological University, Singapore, Singapore
e-mail: jschlundt@ntu.edu.sg

X. Deng et al. (eds.), *Applied Genomics of Foodborne Pathogens*, Food Microbiology and Food Safety, DOI 10.1007/978-3-319-43751-4_2

spread from animals [1]. A number of factors, including poverty, increasing population density, disruption of wildlife habitats, increased food trade and changes in food preservation and consumption habits have resulted in increased risks of contraction of infectious diseases and subsequently their potential global spread. Globally, about 23 % of all deaths are caused by infectious diseases, with the most significant burden in developing countries [2]. Nearly all of the most important human pathogens are either zoonotic or originated as zoonoses [3–6]. Striking examples include HIV/AIDS and Spanish influenza, which started by interspecies transmission of the causative agents [7–10] and have caused millions of deaths worldwide and more recently SARS and MERS coronaviruses and H1N1 and H5N1 influenza A viruses.

Detection and surveillance form the backbone of all systems currently used to control infectious diseases worldwide. However, surveillance is still typically targeted at a relatively limited number of specified diseases, and, maybe more importantly, there is a very significant global disparity in national disease detection systems and methodology. In particular, public health efforts and patient treatment are hampered by different obstacles: the use of different, specialized, expensive and difficult-to-compare detection techniques; a lack of collaboration between different microbiological fields; (inter)national politics on the disclosure of (patient) information and research data; intellectual property rights; and, a lack of sufficient diagnostic capacities particularly in developing countries. A more effective and rational approach to the prevention of microbial threats is essential at the global level. Efforts to mitigate the effects of infectious threats, focusing on improved surveillance and diagnostic capabilities, are crucial [11]. With recent technological advances and declining costs in the next generation sequencing field, these tools will play an increasingly important role in the surveillance and identification of new and previously unrecognized pathogens in both animals and humans but also for identification and characterization of traditional pathogens. Inherently an enormous increase in microbial whole genome sequences (WGS) is to be expected, providing a wealth of information to aggregate, share, mine and use to address global public health and clinical challenges. The goals of the Global Microbial Identifier (GMI) initiative in this respect will be outlined below.

Next Generation Sequencing and Whole Genome Sequencing: A New Potential for Integrated Surveillance of Infectious Diseases

Surveillance is a key component of preparedness for infectious diseases, and is done globally to monitor trends in endemic diseases (e.g. influenza, dengue, salmonellosis), to monitor eradication efforts (polio, measles, brucellosis), or to signal unusual disease activities. Molecular diagnostic tools, which rely on the recognition of short pieces of unique genome sequence (e.g. PCR and microarray (biochip) technologies) and provide sensitive and specific detection and sufficient genetic diversity for subtyping, are used routinely in clinical diagnostic and surveillance settings. Although the partial genome information, such as epidemiological markers, often is sufficient for patient management and basic surveillance objectives, from a public health perspective the increasing capacity for more extensive sequencing most likely will increase the depth of information gathered on pathogens and disease. Recombination and reassortment of viral genomes for instance may generate future threats; influenza A viruses for example are able to undergo reassortment if a single cell is concurrently infected with more than one virus [12]. These reassortment events can dramatically change the evolution of influenza A viruses in a certain host and lead to new epidemics and pandemics. Such events may easily be missed when surveillance is relying on molecular diagnostic tools that target small microbial genome fragments.

Whole genome sequencing (WGS) is a laboratory process that determines the complete genome sequence of an organism under study providing significantly more information than routine molecular diagnostic tools. This can have important implications; for instance during the recent outbreak of MERS coronavirus in the Middle East, analysis of small genome fragments did not provide sufficient phylogenetic signal for reliable typing of virus variants [13]. Classically, whole microbial genome sequences were determined by PCR and Sanger sequencing. Nowadays next generation sequencing (NGS) techniques are used increasingly in the human medical sciences, and are now also widely used to identify and genotype microorganisms in almost any microbial setting [14–17]. There are different NGS techniques targeting single microorganisms or a complete metagenome in a sample through methods unrelated to specific sequence recognition.

A cascade of technological NGS advancements both in the analytical sequencing field (e.g. pyro- and nanopore sequencing) and in the information technological (IT) field (e.g. increasingly faster and cheaper internet, computing rates and storage capacities; and the development of NGS software tools) has decreased the cost of WGS much faster than predicted 10 years ago (Fig. 2.1). Today, the actual cost of sequencing an average bacterial genome of 5 Mb would in practise cost between USD 50–100. It is estimated that both the price and the speed of WGS analyses will decrease to a point where it can seriously compete with traditional routine diagnostic identification techniques. The enormous potential of WGS in the surveillance of infectious diseases [18,19] has been demonstrated in many studies now including the tracking and tracing of the cholera outbreak in Haiti in 2010 [20], the Enterohaemorrhagic *Escherichia coli* (EHEC) outbreak starting in Germany in 2011 [21] and others e.g., [22,23]. During the EHEC outbreak, scientists from around the globe performed NGS and shared their results for analysis. The collaboration between these researchers allowed for joint and rapid analysis of the genomic sequences, revealing important details about the involved new strain of *E. coli*, including why it demonstrated such high virulence. Similar collaborations exist globally during emerging viral infections such as MERS coronavirus. Continuing innovations, however, are required to allow NGS techniques to become standard in clinical practice. In addition, hurdles regarding ethical, legal, social and societal issues need to be overcome.

It seems certain that NGS techniques will play an increasingly important role in the identification of new and previously unrecognized pathogens and inherently a large increase in the total amount of microbial whole genome sequences is to be expected.

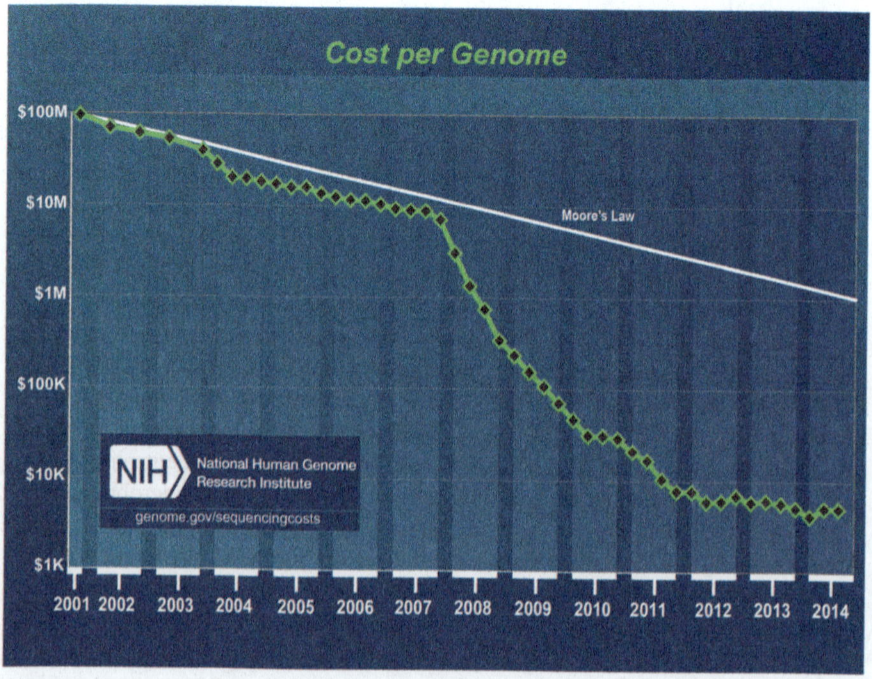

Fig. 2.1 NGS cost per raw magabase of DNA sequence. Taken from the National Human Genome Research Institute (http://www.genome.gov/sequencingcosts/)

As a consequence of the steadily decreasing costs of WGS, an increasing number of microbiological laboratories have embarked on WGS projects to characterize own stocks of infectious agents in their existing biobanks. This in turn generates huge amounts of genomic data in private databases as well as significantly increased numbers of genomes to the global DNA databases such as GenBank. This genomic information is, however, not fully interconnected and in most cases not accompanied with sufficient (national or international) metadata. The need to integrate these databases and to harmonize data collection has been generally recognized by the scientific community for some time [24]. Further integration of these databases and linking the genomic data to metadata for optimal prevention of infectious diseases, and to make it fit for other uses including routine diagnostics, is the new challenge.

Notably, while future use of WGS is likely to boom in developed countries, an even more dramatic change in developing countries creates a potential for a significant diagnostic leap-frog in these countries. While current diagnostic methods are diverse and require a lot of specialized training, NGS holds the potential of a simple one-size-fits-all tool for diagnosis of all infectious diseases, thereby dramatically improving public health in developing countries. At a systemic level, the use of NGS will enable uniform laboratory-, reporting- and surveillance-systems not only relative to human health, but reaching out to the identification of microorganisms in all other habitats, including animals and the environment: a true 'One Health' approach [25]. At the same time the development of new centralized and de-centralized diagnostic systems will be significantly simplified with the potential of real-time characterization of microorganisms in individual, local decentralized labs with sequencers and internet link-up. Recent studies have shown that it is possible to determine the species, type as well as the antimicrobial/antiviral susceptibility of both bacterial and viral pathogens, even when using sequencing directly on clinical samples [18, 26]. This would be even more valuable for clinical laboratories in developing countries that do not currently have the same diagnostic capacities as most developed countries.

As NGS technology spreads more globally, there is an obvious potential to develop a global system of whole microbial genome databases to aggregate, share, mine and use microbiological genomic data, to address global public health and clinical challenges, and most importantly to identify and diagnose infectious diseases. Such a system should be deployed in a manner which promotes equity in access and use of the current technology worldwide, enabling cost-effective improvements in plant, animal, environmental and human health. If the system is set up in an 'open access' format it would likely enable comprehensive utility of NGS in developing countries, since the development of open databases and relevant algorithm platforms at the global level would enable immediate translation of sequence data to microbial identity and antimicrobial resistance pattern. In general, it is necessary to have a comprehensive database of all known microbial DNA sequences to make full use of locally derived DNA sequence to identify and characterize your isolate microbiologically and epidemiologically. A global system, supported by an internationally agreed format and governance system, will benefit those tackling individual problems at the frontline (clinicians, veterinarian, epidemiologists, etc.) as well as other stakeholders (i.e. policy-makers, regulators, industry, etc.). By enabling access to this global resource, a professional response on health threats will be within reach of all countries with (even relatively simple) basic laboratory infrastructure.

The Global Microbial Identifier (GMI) Initiative

The GMI initiative attempts a description of the landscape and opportunities of the global NGS/WGS field and suggests a collaborative effort to bring together different microbiological fields with the purpose of creating a global microbial identifier (GMI) tool on the basis of WGS data. To achieve this, GMI envisions a WGS database and analytic tools that are used and maintained by multidisciplinary researchers, clinical microbiologists, food scientists, (bio)informaticians, veterinarians, physicians, and other stakeholders. This database should be useful for basic research and for identification and disease diagnosis of any possible microorganism. In September 2011 the first international GMI conference was organised in Brussels[1] to discuss the possibility to use WGS as a microbiological diagnostic tool on a global scale [27]. At

[1] Perspectives of a global, real-time microbiological genomic identification system—implications for national and global detection and control of infectious diseases. Consensus report of an expert meeting 1–2 September 2011, Bruxelles, Belgium. Available at http://www.globalmicrobialidentifier.org.

this stage several preconditions for a successful initiation of an initiative of this sort seemed to have been met: (1) WGS had become mature and a potential serious alternative for other genotyping techniques, (2) the price of WGS had been falling dramatically and was now in some cases below the price of traditional methods, (3) vast amounts of IT resources and a fast internet had become available in most parts of the world, and (4) suggestions had been made that a One Health (human/animal) approach could enable improved control of infectious diseases [28].

Currently, GMI organizes annual meetings to discuss progress and future development. These meetings are organised and attended by a number of scientists and policy makers from around the world, including the World Health Organization (WHO), the UN Food and Agricultural Organization (FAO), the World Organisation for Animal Health (OIE), the United States Food and Drug Administration (US FDA), the European Commission (EC), the United States Centers for Disease Control and Prevention (CDC), the European Centre for Disease Prevention and Control (ECDC), the National Food Institute of Denmark, the European Food Safety Authority (EFSA) and several other universities, food research institutes and public health institutions. The general conclusion of the first meeting was that the spread of the WGS technology for microorganisms should be linked to the establishment of a global genomic database for microorganisms. This would entail an interactive, global, open source database supported by scientists from all regions of the world and from all fields, including human health, animal health, food safety and environmental health, and holding information on bacteria, viruses, fungi as well as parasites, together with important metadata relating to host information, environmental factors, sequencing methods, and other microbiological and epidemiological details. The structure and platform of the database(s) should be such that it could be used by different software tools (algorithms etc.) to generate meaningful results from data in the database.

Landscaping the Global Microbial WGS Field

The current steep rise in the potential of NGS has led to several developments around the globe: new fields of science have been strengthened (e.g. bioinformatics and its subfields); established scientific fields utilize NGS in novel ways; new WGS software tools are put online every week; multiple companies offer NGS and WGS equipment and services; and also at governmental level, NGS is considered in the continuous quest for public health efficiency improvements. These developments make NGS grow from a basic research tool into a mature general purpose tool; however, the constant danger is that cross-talk between these separate initiatives wanes in typical silo-fashion and that all technical development takes place in the western world (+ China), which might lead to a strong underuse of the total WGS potential.

While many researchers, clinicians as well as public and animal health professionals have made statements in support of the dramatic new potential, there currently is still no coherent description of the global (diagnostic) landscape of WGS and how it could best take over from the traditional techniques, as well as the potential benefits and costs of such development at a global scale. At the same time it should be realized that the free sharing of genomic data will meet significant obstacles, both from the research, the public health—and the food production communities. Important examples, which can already be envisioned, are: (a) the general reluctance of researchers to share data before publication, (b) the reluctance of governments and institutions to share data when competing interests are in play (e.g. trade, tourism etc.), (c) legal and ethical issues including personal information confidentiality and intellectual property rights [29–31].

There is a need to further analyze this landscape. Such analyses should include identifying all stakeholders and their use of NGS, describing the technical and political needs, characterizing the potential future clinical and public health systems enabled by WGS, and in the process, specifically considering the need for capacity building in this area for developing countries [28]. A number of different scientific fields should be included in the analysis (e.g. public health, food safety and production, animal health, environmental health, bioinformatics, clinical science, biotechnology etc.). Likewise different societal sectors should be considered (e.g. healthcare, food and healthcare industry, agriculture, commerce, as well as developmental economics etc.). A description of existing WGS initiatives within different microbiological specializations (virology, bacteriology, parasitology etc.) will be key to understanding this field, as will be a thorough description of existing and future NGS potential in laboratory settings in developing countries.

GMI the Network

Following the inception of GMI in 2011, GMI has grown as a global network of scientists and other experts committed to improving global infectious disease and food safety prevention using WGS. A charter has been drawn up in which the network partners have agreed on its mission and vision (http://www.globalmicrobialidentifier.org/). In short, the mission is to build a global network for microbiological identification and infectious disease surveillance using an open and interactive worldwide network of databases for standardized identification, characterization and comparison of microorganisms through whole genome sequences of microorganisms. GMI's vision is a world where high quality microbiological genomic information from human, food, animal and plant domains is shared globally to improve public health, healthcare, a healthy environment and safer food for all.

The GMI network essentially is a global network of stakeholders that take part in shaping how the database and its supporting structures can best be defined, set up and used. Figure 2.2 shows a simplified impression of GMI: the GMI users, the database(s), the GMI software pipelines and other analytical tools, and the GMI organization. The users include anybody using the GMI database such as medical and veterinary labs, physicians and veterinarians, public health institutes, food science and industry etc. The GMI database is defined as all the microbiological WGS data and the linked metadata that can be accessed by GMI software. GMI software includes any software tool or software pipeline designed to interact with the GMI database to produce

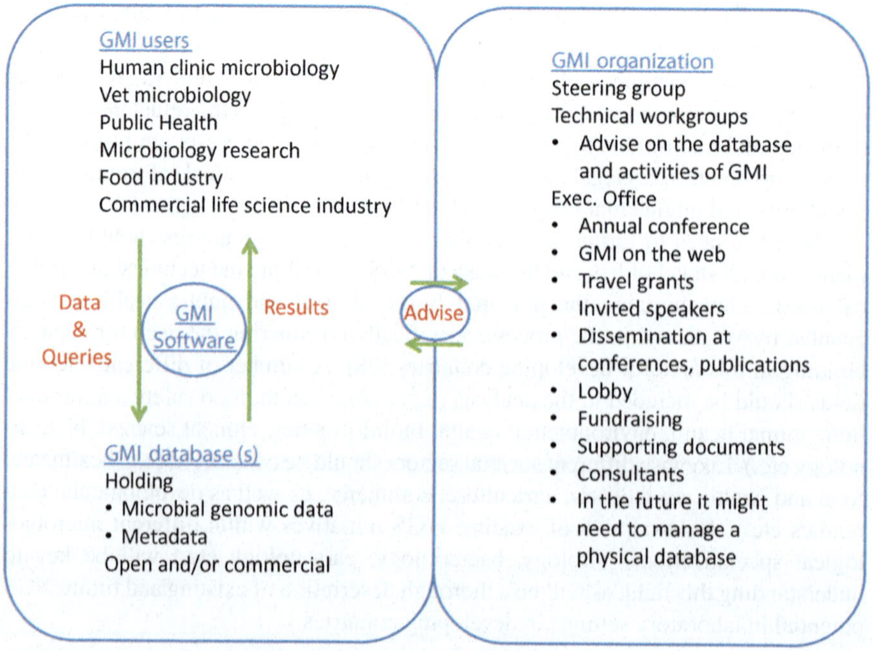

Fig. 2.2 Schematic outline of GMI

results, e.g., genome assembly, data comparison, disease diagnosis, resistance prediction, simple data extraction, data viewing etc. The GMI organization includes the people creating the database(s), people helping the development of the necessary software, and people active in the GMI working groups and steering committee.

GMI is now a global initiative with a defining Charter, annual conferences, a website and regular newsletters.[2] The eighth global GMI conference (GMI8) was held in May 2015 in Beijing, China, and GMI9 took place in FAO in Rome, Italy in May 2016. GMI is organized through a Steering Committee overseeing four working groups and supported by an executive office. The four working groups are: (1) Political challenges, outreach and building a global network; (2) Repository and storage of sequence and meta-data; (3) Analytical hard- & software and (4) Ring trials and quality assurance.

GMI the Database

The proposed GMI database will consist of all the microbiological WGS data, both annotated (including reference strains) and un-annotated, together with the relevant metadata, all to be accessed by GMI software. Questions related to the status,

[2] Homepage: http://www.globalmicrobialidentifier.org/.

separation and encryption of metadata within the database system need further consideration, including international and national political debate. While the ultimate aim might be one database or a federated database system[3] enabling fast identification through the comparison of a new isolate with many existing reference genomes worldwide, this system could be too complicated initially. Given the right tools, however, this technical complication may be neutralized and a federated system may even allow faster identification by parallel computing. The likely and preferable development in this area will depend on many other factors, including the available software and the state of global internet infrastructure.

A global reference database may be supported by additional database(s) to do the follow-up analysis after a first identification has been achieved, and these databases could potentially be located elsewhere. Considering the technical challenge of complete genome assembly, it becomes important to consider which level of (un-assembled) rough data can be input for assembly and analysis with GMI tools? This issue will potentially disappear when more powerful software is developed. Currently, however, these issues are still bottlenecks when quick turnaround of data analysis is demanded.

Compared to a federated system, centralized storage has several advantages. It will be a one stop shop and its openness may be preserved by the government(s) supporting the database. The creation of a centralized system would not prevent the future addition of regional/local databases to the structure to create a federated database system, which may potentially become necessary anyway if the future amount of data becomes too large for a single location. Such additional, federated databases may also be commercial, and this may hold both risks as well as advantages. The key will most likely be that the software adding and retrieving data can reach all relevant information. This means that either the software needs to handle multiple formats used by different databases, or the data structures of different databases should be similar. In addition, commercial databases should address how they can be accessed by GMI software and how users pay for their database use. Clearly this involves many controversial issues. Commercial involvement may on the one hand put limitations on the development of the GMI database and the speed at which it will evolve. On the other hand, in economic terms it will have a great spin-off in terms of companies that may offer services to and depend upon the GMI database, in a way somewhat similar to the functionality of the internet at present. Such spin-off activity may be beneficial for the quality and quantity of the use of the database. Taken together these are all important issues that GMI aims to discuss and solve through the work of its different work groups and through open discussion and interaction with all stakeholders in the field.

GMI the Software

In addition to the database, a proper functioning GMI system needs software. This software could be located as part of the database in a way similar to software offered by NCBI and linked to GenBank, such as BLAST, and other parties may also offer

[3] A system in which several databases appear to function as a single entity.

software that uses the data from the GMI database to generate comparisons and analyses that the GMI community asks for. An example of this could be the ResFinder software from the Center for Genomic Epidemiology at the Technical University of Denmark (see Table 2.2), which can be used to predict antibiotic resistance profiles from WGS data [18]. On the internet there is a wealth of such tools available and new bioinformatics tools are constantly released, either under an open source license or as commercial software packages or services. The current list of tools is very long and includes many different packages able to perform many different analyses. Unfortunately, there is a lack of coordination and awareness among developers and users. Some tools have been a repetition of already developed tools; some have overlapping analyses; and some are simply outdated already when they occur. On the other hand it is a welcome development that the bioinformatics community is flourishing with an abundance of tools, and GMI could take up the task of providing a portal to help users navigate among tools. Table 2.1 provides a list of several of these tools and links to webpages important for the field of WGS microbiology.

The whole field of "analytical tools" is currently developing fast and there are many different and new initiatives. Ideally, there is a need for simplicity and some of the individual tools developed are being sequentially combined into analytic pipelines. However, there is still much effort needed in this field, because not all programs are compatible with each other, some are not user friendly, are not maintained or are only available on specific platforms. GMI work groups 3 and 4 have taken initiative to investigate what is available and what would be necessary to have for a GMI database to function as a general diagnostic tool. Further advances in the software tools should aim at answering specific questions from the different fields of microbiology. Important advances in this area will be to generate more user friendly software to take the tools that now mainly are geared towards the bioinformatics and basic science communities, to the first-line users (clinicians and public health and food safety professionals) e.g. to help the clinical field with disease diagnosis or to help with complicated global tracking and tracing analyses relative to food contamination or infectious diseases. User friendliness would increase the use of the GMI database and thereby its value. It may be envisioned that this may come through the combination of apps and online software tools generated for use on (super) computers down to smart phones. In addition, bringing together different software routines that currently need to be run separately, will contribute to this. The chapter on comparative genomic epidemiology (CGE) elsewhere in the book gives an extensive overview and discussion of CGE tools for WGS microbiology.

Metadata and Depth of Analysis

Metadata is data that describes other data, and in many cases represent data that are necessary to make epidemiological sense of WGS data. Metadata relative to the sequence data of a clinical isolate in the database would for instance be patient demographics, geographical location and method of isolation etc. The more details there

Table 2.1 Short overview of some of the WGS analysis tools found on the internet

Tool	Link	Short description
Online analysis tools	http://molbiol-tools.ca/	Lists numerous bioinformatics tools
VFDB	http://www.mgc.ac.cn/VFs/	This database provides BLAST-based identification of virulence genes in 26 genera of bacterial pathogens. The database aims at being the most comprehensive database of virulence factors and hence also contains, for instance, hypothetical proteins
ResFinder	http://cge.cbs.dtu.dk/services/ResFinder/	ResFinder identifies acquired antimicrobial resistance genes in total or partial sequenced isolates of bacteria
ARDB	http://ardb.cbcb.umd.edu/	A manually curated database (ARDB) unifying most of the publicly available resistance genes and related information. Regular BLAST and RPS-BLAST tools would help the user to identify and annotate new potential resistance genes by blasting against ARDB DNA or protein sequences. Has not been maintained since 2009
BTXpred	http://www.imtech.res.in/raghava/btxpred/	The BTXpred server aims at predicting whether an amino acid sequence is a bacterial toxin or not, whether it is an endo- or exotoxin, and the function of exotoxins. It requires amino acid sequences as input
RASTA-Bacteria	http://genoweb1.irisa.fr/duals/RASTA-Bacteria/	RASTA-Bacteria is aimed at the identification of TA modules (toxins/antitoxin modules)
The comprehensive antibiotic resistance database	http://arpcard.mcmaster.ca/	The RGI provides automated annotation of your DNA sequence(s) based upon the data available in CARD, providing prediction of antibiotic resistance genes
t3db	http://www.t3db.org/	t3db is a database containing toxins and targets along with detailed information collected from various sources. It does not focus solely on bacterial virulence factors, but includes pollutant, pesticides, and drugs. Also, it is very strict with the inclusion of toxins and only includes toxins for which the structure is known
DBETH	http://www.hpppi.iicb.res.in/btox/	DBETH is a database of bacterial exotoxins for humans. As it requires amino acid sequences as input
VICMpred	http://imtech.res.in/raghava/vicmpred/	VICMpred is an SVM-based method for prediction of toxins (and other functional proteins) based on amino acid sequence

(continued)

Table 2.1 (continued)

Tool	Link	Short description
Samtools	http://samtools.sourceforge.net/	SAM Tools provides various utilities for manipulating alignments in the SAM format, including sorting, merging, indexing and generating alignments in a per-position format
Figtree	http://tree.bio.ed.ac.uk/software/figtree/	Tree viewer
Velvet (combined with VelvetOptimiser)	http://bioinformatics.net.au/software.velvetoptimiser.shtml	*de novo* assembler
BWA	http://bio-bwa.sourceforge.net/	Sequence mapper
AdapterRemoval	https://github.com/slindgreen/AdapterRemoval	Trimming and adapter removal from raw read data

are in the metadata the more detailed the tracking and tracing of microorganism can be. However, a higher level of detail can also result in political and/or privacy/ethical complications, especially for the people publishing the data [29]. Without metadata, one would have a 'genotype' database only containing peta- to exabytes of WGS data. This would already be a giant step for mankind as we will discover many new genomes and microbial communities. However, to use genomes for infectious disease investigation and epidemiology, metadata are essential. The list and structure of metadata should be concise and include only what is defined as essential while excluding redundant or unethical information. For instance, making a distinction between men and women, children and the elderly would be very informative and may be essential for clinical data. However, it would be under discussion whether to include race in the list of metadata, even though there might be situations imaginable for which having such metadata would help solve scientific questions. Also, different fields of research or policy making may have use for different metadata. For instance, economists studying the economic cost of a certain disease will be interested in the number of outbreaks and relations between different economically important sectors e.g. specific food or food preparing sectors, while public health specialists and clinicians might be more interested in resistance phenotypes, treatment options etc.

In general, it is thus essential to generate a list of metadata that can be considered essential for each sample. In addition, per discipline this list may be extended with field specific metadata which are essential for each individual field. Also, there should be a list of metadata to be avoided by GMI. Such thinking would bring us roughly three lists of metadata: the minimal essential, the field specific list, and the list of metadata to be excluded. To help the discussion on this one may categorize each of these three lists further into essential and optional data. Table 2.2 gives a very basic and simplified example of how such lists might look like to help the discussions on what these lists should finally comprise. Generating the different field specific list in a collaborative manner as done in GMI will potentially be beneficial

Table 2.2 Example of types of metadata that may be valuable for the GMI database

General metadata	Examples work field specific metadata		Not essential for GMI
	Clinical examples	Food examples	
Essential			
Submitters contact info Submitters identifier Unique GMI identifier Name organism Name strain Alias(es) Date isolated Attribute package (if pathogenic type of pathogen) Isolation source Cultured Yes/No Geographical origin of the sample (Country and City) Lab strain Yes/No Reference strain Yes/No	Host (human/animal) Host sex Host age range Host age Name(s) disease Zoonotic Y/N Resistance profile Treatment options Confirmation tests Confirmation tests results Short (standard) case description Outbreak Y/N, plus code	Specific name source GPS location source Climate type source Location in source depth Zoonotic Y/N Resistance profile Virulence Confirmation tests Confirmation tests results Outbreak Y/N, plus code Commercial source Y/N Producer (kept secret?) Short case description	Race host HIV status host Links to hosts social network profile
Optional			
Colony color Description of the sample/source/strain Detailed geographical origin of the sample (GPS coordinates etc) Growth rate Growth on cell lines	Outbreak code Growth rate Growth on cell lines PFGE pattern codes Serovar Outcome other tests Suggestion for further testing Short (standard) case description	Host (human/animal) Host Sex Host age range Host age Name(s) disease PFGE pattern codes Serovar Growth rate Colony color Growth on cell lines Suggestions for further testing Short (standard) case description	Occupation of host Use in terror attack

These are examples of metadata that different microbiological fields would like to collect and serves to illustrate the concept. Further discussion in GMI will be needed to generate agreed lists of metadata required for each microbiology field, and as the technology progress these list may have to be updated

for the end results, since different sectors would be able to follow each other's progress and may minimize redundancy in the different lists. Furthermore, it should be decided which data should be collected but kept confidential and only accessible by the submitter and others with permission of the submitters. For instance, should it be open source information if a specific producer is linked to a specific microorganism and/or outbreak, or should such data be managed (and kept secret or open) by the relevant regulatory agencies?

There are a number of technical questions adding to the complexity of the issue. Should reference strain data have a different set of metadata than the metadata required when submitting and/or comparing one's samples to the GMI database? And, where to best store all the metadata? The different types of metadata originating from different fields might be centrally stored which will have advantages for retrieving and working with them, and for ensuring the open source character of the database. However, this is not necessary and by using the right identifiers, different metadata databases may be generated that connect to a central WGS database. Central storage of all kinds of metadata may bring the advantage of having a one stop shop for everything, and it may help to find new cross links between data from different fields. It may, however, also lead to confusion when users are overloaded with too much information that is not necessary for their purpose.

Quality Assurance and Testing

Investigating whether GMI users will be able to perform DNA extraction, library preparation, the actual sequencing, the assembly and phylogenetic analysis following different laboratory protocols, software tools, and sequence platforms will enable an evaluation of the reliability of submitted sequence data to a GMI database [32]. GMI aims to assist laboratories and partners globally to perform NGS to the highest quality level, and to prepare for this GMI in 2013 conducted a survey to identify the intended end-users, priority organisms, and quality markers for proficiency testing [33]. GMI in 2014 performed a pilot proficiency test with a limited number of laboratories to test the developed IT system and corresponding protocol. The GMI 2015 proficiency test was fully rolled out by December 15 (supported by the EU/COMPARE programme (www.compare-europe.eu) and the USFDA GenomeTrakr and Microbiologics®. This first global proficiency test in this area had a focus on *Salmonella enterica*, *Escherichia coli* and *Staphylococcus aureus*, and allowed for sign-up for each species separately (see www.globalmicrobialidentifier.org). 55 laboratories, from all continents, signed up for the test. The main objective of this proficiency test was to assess the feasibility of achieving reliable laboratory results of consistently good quality within the area of DNA preparation, sequencing, and analysis (e.g. for the use relative to phylogeny, MLST, resistance genes etc.). This will in time ensure or enable harmonization and standardization of whole genome sequencing and data analysis, with the final aim to produce comparable data for the GMI initiative, and thereby consistent data for the GMI database. A further objective is to assess and improve the data uploaded to databases such as NCBI, EBI and DDBJ. Therefore, the laboratory analysis work performed for this type of proficiency testing should be done employing the methods routinely used in the individual laboratories. The proficiency testing performed in this area has consisted of two wet-lab and one dry-lab components targeting *Salmonella*, *E. coli* and *S. aureus*. The wet-lab components assess the laboratories' ability to perform DNA preparation, sequencing procedures and analysis of epidemiological markers whereas the dry component assesses the ability to analyse a whole-genome-sequencing dataset and distinguish between clonally related and sporadic genomes. At present (September 2016) the GMI2016 proficiency testing is ongoing. The future vision of the proficiency testing is to target lower priority bacterial pathogens as well as to develop a parallel proficiency testing regime targeting viruses (http://www.globalmicrobialidentifier.org/News-and-Events/Previous-meetings/7th-Meeting-on-GMI).

Other laboratory methods are being discussed and optimized for use in GMI, including consideration of how to include other types of microorganisms in proficiency tests and how to initiate parallel viral pilot proficiency test schemes including RNA methods.

Next to quality assurance at the laboratory level it will be important to have a reliable source of analytical tools that cover the different tasks requested by the GMI users and are of sufficient quality to be used in different (sometimes more sensitive) settings than basic research, for instance in clinical settings. GMI aims to define the functional requirements for these tools from the perspective of end-users (clinical, public health, research) in terms of applications needed (identification, outbreak detection etc.) and priority microorganisms and diseases. To do so, GMI maps the currently available analytical software tools as well as developments in the field and benchmark them against the needs of GMI end-users in order to identify implementation gaps and projects that may fill those gaps. By this mapping effort and through software testing, GMI aims to construct a central portal of tools, to indicate a quality level, and state the usefulness and the user friendliness of the different tools for the different GMI end-users. Through this effort it will be possible to provide guidance for further development of (new) analytical tools.

In addition to the development of these testing schemes, which will get a more permanent shape and place in the future GMI network, GMI plans to design *in silico* pilots using realistic scenarios based on and using data from a previous infectious disease outbreak or another event (http://www.globalmicrobialidentifier.org/News-and-Events/Previous-meetings/7th-Meeting-on-GMI). The goal of these pilots will be to help shape the process and the form that the GMI tools take, develop training skills and increase the participation level. In particular, this would be important to increase the participation level of members that currently lack the necessary laboratory capacity including members from many developing countries. These pilots will address several important issues and may help answer some important questions such as: How well does data transfer work? How well does data analysis, including species identification and outbreak clustering, work? What are the biggest challenges for coordinating an analysis that is highly dependent on metadata? What are the minimum standards required to run the system? And finally, what might be the gain in turn-around time?

Concluding Remarks

Several already existing internet-based genomic tools and databases have been presented and discussed at GMI global meetings—all of which are generally aimed at improving (inter)national detection and identification of different types of microorganisms. For instance, the global programme PulseNet compares the PFGE 'DNA fingerprints' of bacteria from patients to find clusters of disease that might represent unrecognized outbreaks (http://www.cdc.gov/pulsenet/). MLST-net can be used to compare various bacteria on the basis of multilocus sequence typing (MLST) analysis (http://www.mlst.net/). EuPathDB (http://eupathdb.org/eupathdb/) and ZoopNet [34] are portals for accessing genomic-scale and MLST datasets, respectively, which are associated with eukaryotic parasites. And NoroNet is a network of public health institutes and universities sharing virological, epidemiological and molecular data on norovirus and includes a tool for Norovirus identification and epidemiology on the basis of sequence comparison (http://www.rivm.nl/en/Topics/N/NoroNet). In contrast to GMI these earlier networks had to focus their effort on a single technique and often a limited group of microorganisms to make comparisons possible. With the arrival of cost-effective NGS and WGS this is no longer necessary; the different microbiological fields may now work together and different WGS analytical tools can be exchanged to maximize efficiency. Many of these earlier networks are now trying to make the move from traditional techniques to NGS. For example, PulseNet investigates how to use WGS and potentially metagenomics to replace PFGE and thus have a culture independent and faster technique (see: http://www.cdc.gov/pulsenet/next-generation.html). In such transitions it is important to implement new techniques in a way such that the old and new techniques are comparable and no data are lost.

GMI is an initiative open to anyone interested and many of the people associated with the networks summarized above have actually participated in the initiation and development of GMI. The work of GMI is to promote inter-disciplinary and international discussion of potential synergistic solutions to optimize the use of WGS globally. This process will take time and although some work may progress quickly (e.g. proficiency testing) for other issues more time is needed, as is inter-governmental debate and agreement. The roadmap for the development of the database that has been proposed with a vision of constructing an international system by 2020 is as follows:

- Development of pilot systems.
- Initiation of appropriate 'legal entity', with the formation of an international core group and governance structure
- Analysis of the present and future landscape to build the database
- Diplomacy efforts to bring the relevant groups together
- Development of a robust IT-backbone for the database
- Development of novel genome analysis algorithms and software
- Construction of a global solution, including the creation of networks and regional hubs

Initiatives with similar or overlapping goals as GMI have emerged and should be used to explore the opportunity for collaboration and synergy. Examples of such initiatives are the global alliance for genomics and health (http://genomicsandhealth.org/) and that of Google Genomics (https://developers.google.com/genomics/), both mainly focusing on human genomics, and the initiatives of CDC to use

WGS in parallel to their PFGE diagnostics and in their AMD programme (Advanced Molecular Detection) as well as the creation of the USFDA Genome Trackr Network, linking public health and university laboratories that collect and share genomic and geographic data from foodborne pathogens. GMI is presently in contact with these initiatives in order to investigate the potential for collaboration and synergy in the area of NGS/WGS use in microbiological identification and research as well as in genomic epidemiology and food microbiology.

References

1. Jones KE, Patel NG, Levy MA, Storeygard A, Balk D, Gittleman P, Daszak P. Global trends in emerging infectious diseases. Nature. 2008;451:990–3.
2. Mathers CD, Boerma T, Ma Fat D. Global and regional causes of deaths. Br Med Bull. 2009;92(1):7–32. doi:10.1093/bmb/lpd028.
3. Kuiken T, Leighton FA, Fouchier RAM, et al. Pathogen surveillance in animals. Science. 2005;309(5741):1680–1.
4. Smith GJD, Vijaykrishna D, Bahl J, et al. Origins and evolutionary genomics of the 2009 swine-origin H1N1 influenza A epidemic. Nature. 2009;459(7250):1122–5.
5. Taylor LH, Latham SM, Mark EJ. Risk factors for human disease emergence. Philos Trans R Soc Lond B Biol Sci. 2001;356(1411):983–9.
6. Woolhouse MEJ, Gowtage-Sequeria S. Host range and emerging and reemerging pathogens. Emerg Infect Dis. 2005;11(12):1842–7.
7. de Wit E, Kawaoka Y, de Jong MD, Fouchier RAM. Pathogenicity of highly pathogenic avian influenza virus in mammals. Vaccine. 2008;26 Suppl 4:D54–8.
8. Gao F, Bailes E, Robertson DL, et al. Origin of HIV-1 in the chimpanzee Pan troglodytes troglodytes. Nature. 1999;397(6718):436–41.
9. Hirsch VM, Olmsted RA, Murphey-Corb M, et al. An African primate lentivirus (SIV) closely related to HIV-2. Nature. 1989;339:389–92.
10. Osterhaus A. Catastrophes after crossing species barriers. Philos Trans R Soc Lond B Biol Sci. 2001;356(1410):791–3.
11. Osterhaus ADME, Smits SL. Genomics and (Re-) emerging viral infections. In: Ginsburg GS, Willard HF, editors. Genomic and personalized medicine. 2nd ed. Amsterdam: Elsevier; 2012. doi:10.1016/B978-0-12-382227-7.00097-5.
12. Steel J, Louwen AC. Influenza A virus reassortment. Curr Top Microbiol Immunol. 2014;385:377–401.
13. Smits SL, Raj VS, Pas SD, Reusken CB, Mohran K, Farag EA, Al-Romaihi HE, AlHajri MM, Haagmans BL, Koopmans MP. Reliable typing of MERS-CoV variants with a small genome fragment. J Clin Virol. 2015;64:83–7. doi:10.1016/j.jcv.2014.12.006.
14. Liu GE. Recent applications of DNA sequencing technologies in food, nutrition and agriculture. Recent Pat Food Nutr Agric. 2011;3(3):187–95.
15. Metzker ML. Sequencing technologies—the next generation. Nat Rev Genet. 2010;11(1):31–46.
16. Rogers GB, Bruce KD. Next-generation sequencing in the analysis of human microbiota: essential considerations for clinical application. Mol Diagn Ther. 2010;14(6):343–50.
17. Smits SL, Osterhaus ADME. Virus discovery: one step beyond. Curr Opin Virol. 2013;3:1–6. doi:10.1016/j.coviro.2013.03.007.
18. Hasman H, Saputra D, Sicheritz-Ponten T, Lund O, Svendsen CA, Frimodt-Møller N, Aarestrup FM. Rapid whole-genome sequencing for detection and characterization of microorganisms directly from clinical samples. J Clin Microbiol. 2014;52(1):139–46. doi:10.1128/JCM.02452-13.

19. Reuter S, Ellington MJ, Cartwright EJ, Köser CU, Török ME, Gouliouris T, Harris SR, Brown NM, Holden MT, Quail M, Parkhill J, Smith GP, Bentley SD, Peacock SJ. Rapid bacterial whole-genome sequencing to enhance diagnostic and public health microbiology. JAMA Intern Med. 2013;173(15):1397–404. doi:10.1001/jamainternmed.2013.7734.
20. Hendriksen RS, Price LB, Schupp JM, Gillece JD, Kaas RS, Engelthaler DM, Bortolaia V, Pearson T, Waters AE, Upadhyay BP, Shrestha SD, Adhikari S, Shakya G, Keim PS, Aarestrup FM. Population genetics of *Vibrio cholerae* from Nepal in 2010: evidence on the origin of the Haitian outbreak. MBio. 2011;2(4), e00157-11.
21. Mellmann A, Harmsen D, Cummings CA, Zentz EB, Leopold SR, Rico A, Prior K, Szczepanowski R, Ji Y, Zhang W, McLaughlin SF, Henkhaus JK, Leopold B, Bielaszewska M, Prager R, Brzoska PM, Moore RL, Guenther S, Rothberg JM, Karch H. Prospective genomic characterization of the German enterohemorrhagic *Escherichia coli* O104:H4 outbreak by rapid next generation sequencing technology. PLoS One. 2011;6(7), e22751. doi:10.1371/journal.pone.0022751.
22. Allard MW, Luo Y, Strain E, Li C, Keys CE, Son I, Stones R, Musser SM, Brown EW. High resolution clustering of *Salmonella enterica* serovar Montevideo strains using a next-generation sequencing approach. BMC Genomics. 2012;13:32. doi:10.1186/1471-2164-13-32.
23. Potron A, Kalpoe J, Poirel L, Nordmann P. European dissemination of a single OXA-48-producing *Klebsiella pneumoniae* clone. Clin Microbiol Infect. 2011;17(12):E24–6. doi:10.1111/j.1469-0691.2011.03669.x.
24. Brown EW, Detter C, Gerner-Smidt P, Gilmour MW, Harmsen D, Hendriksen RS, Hewson R, Heymann DL, Johansson K, Ijaz K, Keim PS, Koopmans M, Kroneman A, Wong DLF, Lund O, Palm D, Sawanpanyalert P, Sobel J, Schlundt J, Aarestrup FM. Integrating genome-based informatics to modernize global disease monitoring, information sharing, and response. Emerg Infect Dis. 2012;18(11), e1. doi:10.3201/eid/1811.120453.
25. Wielinga PR, Schlundt J. One health and food safety. In: Yamada A, Kahn LH, Kaplan B, Monath TP, Woodall J, editors. Confronting emerging zoonoses: the one health paradigm hardcover. New York, NY: Springer; 2014.
26. Prachayangprecha S, Schapendonk CM, Koopmans MP, Osterhaus AD, Schürch AC, Pas SD, van der Eijk AA, Poovorawan Y, Haagmans BL, Smits SL. Exploring the potential of next-generation sequencing in detection of respiratory viruses. J Clin Microbiol. 2014;52(10):3722–30. doi:10.1128/JCM.01641-14.
27. Kupferschmidt K. Epidemiology. Outbreak detectives embrace the genome era. Science. 2011;333(6051):1818–9. doi:10.1126/science.333.6051.1818.
28. Schlundt J. The time is right for a global genomic database for microorganisms. Health Dipl Monit. 2012;3(2):2–3.
29. Heger M. Next-gen sequencing shows promise for public health, but faces technical, political, socialhurdles.2011.http://www.genomeweb.com/sequencing/next-gen-sequencing-shows-promise-public-health-faces-technical-political-social.
30. Pisani E, AbouZahr C. Sharing health data: good intentions are not enough. Bull World Health Organ. 2010;88(6):462–6. doi:10.2471/BLT.09.074393.
31. Roche DG, Lanfear R, Binning SA, Haff TM, Schwanz LE, Cain KE, Kokko H, Jennions MD, Kruuk LE. Troubleshooting public data archiving: suggestions to increase participation. PLoS Biol. 2014;12(1), e1001779. doi:10.1371/journal.pbio.1001779.
32. Gargis AS, Kalman L, Berry MW, Bick DP, Dimmock DP, Hambuch T, Lu F, Lyon E, Voelkerding KV, Zehnbauer BA, Agarwala R, Bennett SF, Chen B, Chin EL, Compton JG, Das S, Farkas DH, Ferber MJ, Funke BH, Furtado MR, Ganova-Raeva LM, Geigenmüller U, Gunselman SJ, Hegde MR, Johnson PL, Kasarskis A, Kulkarni S, Lenk T, Liu CS, Manion M, Manolio TA, Mardis ER, Merker JD, Rajeevan MS, Reese MG, Rehm HL, Simen BB, Yeakley JM, Zook JM, Lubin IM. Assuring the quality of next-generation sequencing in clinical laboratory practice. Nat Biotechnol. 2012;30(11):1033–6. doi:10.1038/nbt.2403.
33. Moran-Gilad J, Sintchenko V, Pedersen SK, Wolfgang WJ, Pettengill J, Strain E, Hendriksen RS. Proficiency testing for bacterial whole genome sequencing: an end-user survey of current

capabilities, requirements and priorities. Global Microbial Identifier initiative's Working Group 4 (GMI-WG4). BMC Infect Dis. 2015;15:174. doi:10.1186/s12879-015-0902-3.
34. Wielinga PR, de Vries A, van der Goot TH, Mank T, Mars MH, Kortbeek LM, van der Giessen JW. Molecular epidemiology of *Cryptosporidium* in humans and cattle in The Netherlands. Int J Parasitol. 2008;38(7):809–17.

Chapter 3
The Use of Whole Genome Sequencing for Surveillance of Enteric Organisms by United States Public Health Laboratories

David Boxrud and William J. Wolfgang

Introduction

Enteric pathogens are a major source of human illness causing an estimated 9.4 million episodes of foodborne illness, 55,961 hospitalizations, and 1351 deaths [1] in the US annually. Identification of the sources of foodborne illness is important in order to implement measures to control and prevent future cases of disease.

WGS of microorganisms is an advancing technology that has the ability to revolutionize foodborne disease surveillance. WGS has been used previously in a variety of ways including tracking a nosocomial outbreak of carbapenem-resistant *Klebsiella pneumoniae* [2], identifying an outbreak of *Mycobacteria tuberculosis* over a 21 year period [3], tracking an outbreak of methicillin-resistant *Staphylococcus aureus* in a neonatal unit [4], performing molecular surveillance on 2009 H1 influenza [5], and characterization of Ebola virus [6, 7].

Whole genome sequencing (WGS) will transform how the public health system performs foodborne disease surveillance. In the future, WGS will be performed on all PulseNet organisms in order to serotype, subtype, identify virulence markers and identify antibiotic resistance mechanisms. WGS will efficiently perform several critical tasks necessary for foodborne disease surveillance. This approach will likely reduce cost while increasing the ability of Public Health Laboratories (PHLs) to identify foodborne outbreaks. This chapter examines current practices, benefits, and challenges of implementing WGS for surveillance of enteric organisms in US public health laboratories.

D. Boxrud (✉)
Minnesota Department of Health, St. Paul, MN, USA
e-mail: dave.boxrud@state.mn

W.J. Wolfgang
Wadsworth Center/New York State Department of Health, Albany, NY, USA
e-mail: william.wolfgang@health.ny.gov

X. Deng et al. (eds.), *Applied Genomics of Foodborne Pathogens*, Food Microbiology and Food Safety, DOI 10.1007/978-3-319-43751-4_3

Traditional Foodborne Disease Subtyping Methods

One goal of enteric pathogen surveillance is to identify groups of enteric organisms with a common type that indicates that the organisms may have a common origin. Once pathogens are identified that are likely to have originated from a common source, the cases can be interviewed to attempt to determine the source of the illness.

Historically, phenotypic methods have been used to type enteric pathogens. *Salmonella* serotyping is the most common phenotypic typing method, however many other phenotypic methods have been used including biotyping, antibiotic susceptibility testing and phage typing [8]. While these phenotypic methods yield important information about the agents, they often yield inadequate diversity to aid in outbreak investigations and are not useful as a robust typing method.

Genotypic typing methods have been used with varying levels of success. Multiple-locus variable-number tandem repeat analysis [9], Ribotyping [10, 11], plasmid typing [8, 12], array hybridization, and Polymerase Chain Reaction fingerprinting methods such as Amplified Fragment Length Polymorphism [13, 14] have also been used to type enteric organisms. All of these methods have challenges that make them sub-optimal as a typing method. The most common issue is the lack of subtype discrimination necessary to resolve isolates involved in an outbreak from the sporadic isolates. Additionally, some methods have a lack of subtype stability, an inability to type all isolates, and poor epidemiological concordance.

The inception of PulseNet revolutionized enteric pathogen subtyping. PulseNet utilizes Pulsed-Field Gel Electrophoresis (PFGE) in a standardized process throughout a network of participating laboratories [15, 16]. The standardization coupled with improved cluster investigation on a local and national level in the US and Canada has allowed the detection of many local and national outbreaks including *Listeria monocytogenes* in cantaloupes [17], *Salmonella* Typhimurium in peanut butter [18], and *E. coli* 0157:H7 in ground beef in Colorado, US [19].

Despite the successes that PulseNet has had using PFGE, PFGE does have drawbacks. PFGE is difficult to standardize between laboratories, is relatively slow, and is labor-intensive. In addition, some pathogens such as *Salmonella* Enteritidis (*S.* Enteritidis) are very clonal by PFGE, which makes it difficult to identify cases that have come from a common source. Approximately 75 % of the *S.* Enteritidis organisms in the PulseNet national database are comprised of the four most common PFGE patterns (US Centers for Disease Control and Prevention (US CDC), personal communication). For *S.* Enteritidis and other clonal organisms, WGS offers the promise of higher discriminatory power compared to PFGE [20–29] (see case study 1).

In addition to providing low diversity for some organisms, there are also instances where PFGE provides too much discrimination within an outbreak. This phenomenon is illustrated in a 2002 *E. coli* O157:H7 outbreak due to contaminated ground beef in Colorado, US [30]. The initial investigation focused on a single PFGE pattern that was identified in several clinical cases. The investigation identified ground beef from a specific producer as the likely source. Testing of the ground beef revealed several closely related PFGE patterns in the product. Cases with indistinguishable PFGE patterns are investigated as possibly epidemiologically related. In this outbreak, the case definition prevented identification of additional cases which may have been critical to more rapid identification of this outbreak. In *S.* Enteritidis, closely related PFGE types may be harbored in a single WGS cluster (see case study 2) [27]. In case study 2, using the genomic type generated by WGS would allow greater clarity of subtypes to include at the outset of the investigation. These examples illustrate that PFGE does not always have good concordance with epidemiological information. A method that could better identify isolates that come from a common source would greatly benefit more rapid detection of outbreaks.

The desired typing method for outbreak identification would have the following qualities: stability of the pattern over a short period of time, typeability (all isolates should be able to be typed), yield a high amount of discrimination, have a high amount of epidemiological concordance (agree with available epidemiology data) and to reproducibly assign the same type when an isolate is tested multiple times [31]. Additional convenience criteria include: the method should be flexible enough to type different species and strains, should be rapid, and reagents and equipment should not be overly expensive. Additionally, the method should be easy to use and the results should be easy to interpret, total test cost should be reasonable, and the method should be amenable to computerized analysis and incorporation into electronic databases. For a method to be used by multiple facilities as a national network the results must be able to be obtained and analyzed by multiple facilities and still yield consistent results. Early results from retrospective and prospective studies indicate WGS typing will meet most of these criteria (Jones et al. unpublished data; [24, 25].

WGS as a Subtyping Method for Surveillance

As a prelude to adopting WGS for foodborne surveillance, retrospective studies have demonstrated its utility in terms of stability, typeability, discriminatory power, and most importantly, epidemiological concordance in important foodborne disease pathogens including *S.* Enteritidis [27, 29, 32], *E. coli* O157 [33–35], *S.* Newport [36], *S.* Typhimurium [37], *S.* Montevideo [21, 23], *E. coli* 026 [38], and *Listeria monocytogenes* [39]. These studies are invaluable to our understanding of WGS data from actual outbreaks. Some of these studies helped to identify previously unidentified routes of transmission as well as provide information on whether temporally associated outbreaks are from the same source.

A significant challenge for any subtyping method is using the method in combination with case exposure data to prospectively identify outbreaks. Achieving prospective identification of outbreaks ensures that the method has high epidemiological relevance and is also practical to perform in real-time. WGS is beginning to be used prospectively for selected species and serotypes to identify outbreaks. In the US, the CDC and some PHLs have began routinely performing WGS in addition to PFGE in real-time on *L. monocytogenes* in 2013. Because all *L. monocytogenes* cases in the

US are interviewed with a standard, in-depth interview form as part of the *Listeria* Initiative [40], it is expected that this study will be highly informative with regard to epidemiological concordance. Indeed preliminary data has already shown WGS to be more effective at identification of outbreaks than PFGE by identifying more outbreaks from fewer cases (US CDC, personal communication). In Denmark, WGS was also performed prospectively on 46 verocytotoxin-producing *E. coli* in 2012 [28]. The authors concluded that WGS produced results that were in agreement with epidemiology and produced results faster and at a lower cost compared to traditional methods. Additional prospective studies are needed to demonstrate the utility of WGS to identify enteric outbreaks in a timely and cost effective manner in the PHL setting. Currently, several US states (New York (NY), Wisconsin, Minnesota (MN), Washington) are collaborating on a prospective analysis project of *S.* Enteritidis. The participating states are performing WGS at their facility and the sequence analysis is performed at New York's Wadsworth Center. These ongoing studies are expected to inform the development of a national surveillance system.

Current Workflow for WGS-Based Surveillance in Wadsworth Center/NY State Dept. of Health

At the Wadsworth Center/New York State Department of Health (WCNYSDOH), there are three principal WGS projects which each employ different work flows, to sequence enteric organisms; (1) An in house project to sequence all *S.* Enteritidis isolates in real-time; (2) A US CDC supported initiative to sequence selected enteric organisms in real-time; (3) A US Food and Drug Administration (FDA) supported "GenomeTrakr" project to sequence historical environmental isolates to build a national and international reference database (http://www.fda.gov/Food/FoodScienceResearch/WholeGenomeSequencingProgramWGS/ucm363134.htm). Currently, all sequencing is performed in house on the Illumina MiSeq™ sequencer platform and sequence data is streamed to Illumina's BaseSpace™. The three projects, as well as other bacterial and viral samples, are mixed on sequencing runs to optimize multiplexing (see below for discussion). Notably, all sequence data from these three projects and a subset of the associated metadata is made publicly available at National Center for Biotechnology Information (NCBI) in as close to real-time as is possible.

In the first project, since the fall of 2013 all clinical *S.* Enteritidis received through the WCNYSDOH PulseNet laboratory have been sequenced in real-time. Samples are accessioned through our PulseNet laboratory where PFGE is performed and PFGE patterns are uploaded to the PulseNet National Database. Once PFGE patterns have been assigned by the US CDC, the WCNYSDOH PulseNet laboratory releases all *S.* Enteritidis to the Enteric Genomics laboratory through an in house designed and supported Clinical Laboratory Information System (CLIMS). Because sequence data will be made public in real-time, to assure confidentiality, these samples are given anonymous IDs by the CLIMS. Samples are sequenced in batches once a week. DNA is extracted using the Qiagen Blood and Tissue kit™ and the

DNA submitted to the Wadsworth Center Applied Genomics Technology Core for sequencing. At the Core, sample libraries are prepared using the Nextera XT DNA Library Preparation™ kit and combined with other samples to multiplex between 16 and 20 bacterial samples on a run. Sequences are streamed to BaseSpace™ and shared with the appropriate parties through BaseSpace™. Once the run is completed, samples are checked for quality based on US CDC developed standard operating procedure (SOPs) (current quality metrics include 30× average depth for *Salmonella* sp., Q30 ≥ 75 %, >75 % of clusters pass filter). If quality metrics are not met the deoxyribonucleic acid (DNA) sample is re-sequenced. Metadata and its associated sequence data are submitted directly to NCBI (https://submit.ncbi.nlm.nih.gov/subs/sra/) and shortly thereafter become publicly available. The accession numbers and metadata are manually confirmed once the Biosample and sequence reads are released at NCBI. Biosample and Sequence Read Archive (SRA) accessions numbers are recorded in CLIMS.

For these studies, data is analyzed using an in house pipeline that performs reference based high-quality single nucleotide polymorphism (SNP) analysis. The output from the pipeline is then used to identify either new clusters, composed of as few as two isolates, or additions to existing named clusters. Groups of isolates falling within 0–6 SNPs of one another are designated a cluster number (GC-XXX). The clusters IDs as well as sequencing statistics and methods are recorded for each sample in CLIMS.

The 0–6 SNP diversity cut-off for a cluster was obtained through retrospective analysis of 10 epidemiologically defined single source outbreaks from WCNYSDOH [27] and MN PHL [29] in which there was a maximum of three SNP differences within an outbreak (see also case studies below). However, it is expected that this number will be refined as laboratories gain practical experience with WGS as a typing method and as different methods of identifying SNPs are utilized.

In the second study through the US CDC Advanced Molecular Detection (AMD) initiative, all clinical isolates of *Listeria monocytogenes* and non-O157 Shiga-toxin producing *E. coli* received through the WCNYSDOH PulseNet laboratory are sequenced. In addition, sequencing is performed on sample requested by the US CDC for *Salmonella* and *E. coli* O157. For these samples, the US CDC creates an anonymized ID. Samples are extracted and DNA sequenced in real-time and are multiplexed with other samples. Once the sequence data has been streamed to BaseSpace, Quality Assurance and Quality Control (QC/QA) is performed following the US CDC SOP. For samples that pass quality metrics, a biosample accession number is requested from NCBI. Once the metadata associated with the biosample accession at NCBI is confirmed, the Biosample # is uploaded to the PulseNet national database by the Wadsworth PulseNet laboratory. The US CDC is responsible for uploading the sequence data from BaseSpace to NCBI. The US CDC also performs sequence analysis and returns either SNP based trees or Whole Genome Multilocus Sequence Typing ($_{WG}$MLST) trees for relevant cluster investigations.

In the third project, historical environmental enteric isolates are sequenced for the US FDA GenomeTrakr project. These isolates are selected either from the WCNYSDOH repository or are solicited from partners. Upon request, an anonymized

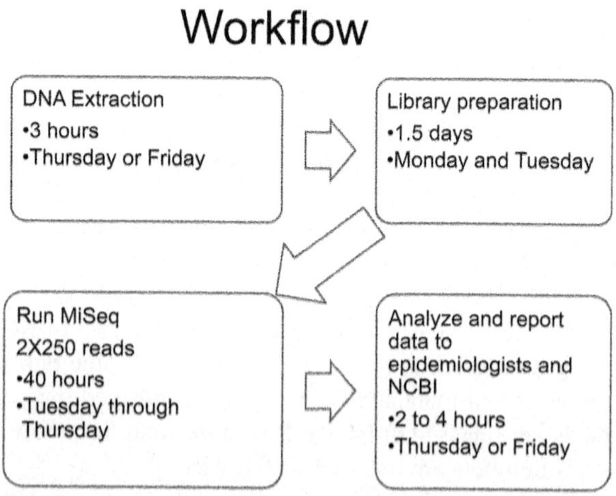

Fig. 3.1 Workflow schematic diagram. Activity is indicated at the top of each box with underlying bullets indicating length of time and usual day of the week for the activity

isolate ID is generated by the US FDA and the metadata is submitted for these isolates to the US FDA. Based on the metadata, the US FDA approves sequencing of the isolates to ensure temporal, spatial, and subtype diversity in the GenomeTrakr database. The US FDA then registers the biosample accessions for these isolates. The biosample accession is returned and added to the MiSeq sequencing worksheet. DNA is extracted for isolates well in advance of the sequencing run. This allows us additional flexibility to fill out sequencing runs of our real-time projects with GenomeTrakr samples when the numbers fall below 16–20 isolates, permitting efficient use of the sequencing reagents. MiSeq runs and projects are shared through Base Space with the US FDA. An automated pipeline at the US FDA performs QA/QC and uploads sequence and metadata to NCBI. Data analysis is performed in the NCBI pathogen pipeline (http://www.ncbi.nlm.nih.gov/projects/pathogens/).

Currently, turn-around time from DNA extraction to analysis is about 1 week (Fig. 3.1). Performing library preparation work immediately after DNA extraction can shorten the turn-around time such that samples submitted on Monday could be reported on Thursday or Friday. A further decrease in turn-around time of 16 h can be achieved by switching from 2 × 250 Illumina read kits to 2 × 150 reads kits, however, this would result in a concomitant decrease in the output so may not be suitable for some applications.

Additional factors such as time from collection to accessioning, re-isolation, and characterization in the laboratory (at the WCNYSDOH PFGE analysis is completed and report to US CDC prior to DNA extraction for WGS, and batching all add substantially to turn-around time. Thus, in the real world, turn-around times will be considerably longer but would be expected to decrease as the system matures. This could be achieved by batching two or three runs each week and eliminating the upfront PFGE analysis and proceeding directly to WGS after accessioning.

Three WGS Case Studies

Case Study 1, a Retrospective Study that Resolved a Cluster in an Endemic PFGE Type

In September of 2010, the Connecticut Department of Health, US identified an outbreak of *S*. Enteritidis affecting a number of residents of a long-term care facility (LTCF) in Fairfield County ("*Salmonella* Enteritidis Outbreak: Long Term Care Facility as Sentinel for a Community Outbreak" http://www.ct.gov/dph/lib/dph/infectious_diseases/ctepinews/vol31no10.pdf). Additionally, there was an epidemiological link to a pastry from a NY bakery that was shared with a number of residents and was the putative source of the infection, though the pastry was never confirmed as the source for the illnesses. Three long-term care facility residents and four NY state residents tested positive for *S*. Enteritidis which had identical PFGE patterns (JEGX01.0004) that were epidemiologically linked (Table 3.1, indicated in 'Epidemiologically Linked' column).

This cohort was chosen as a proof of principle investigation because the PFGE subtype, JEGX01.0004, is seen in about 50 % of all *S*. Enteritidis typed in NY and in the US PulseNet databases. As such, this PFGE type is generally uninformative unless there is a marked increased frequency of appearance.

All *S*. Enteritidis isolate with PFGE pattern JEGX01.0004 s received at the WCNYSDOH from August through October of 2010 (plus some additional isolates from 2011) were selected for sequencing. The DNA from 36 isolates were extracted and run in house on an Ion Torrent sequencer (samples are listed by date and county of isolation Table 3.1). High quality SNP based phylogenetic trees were created as described by Den Bakker et al. [27]. All seven epidemiologically linked isolates were in a single well supported clade with less than 1 SNP average difference (range 0–3 SNPs) between isolates in the clade [27]. Importantly, the analysis revealed nine additional patient samples that were part of the outbreak clade, with a diversity of 0–3 SNPs (indicated in WGS related column). These additional samples were collected in the surrounding community at the same time as the outbreak in the LTCF. Had this information been available at the time of the outbreak it would have potentially aided in traceback investigation to identify the source.

This study illustrated that WGS allowed resolution of genomic clusters within an endemic PFGE pattern allowing greater resolution and significantly improved cluster detection.

Case Study 2 a Near Real-Time Study that Revealed Multiple PFGE Types in a Single Genomic Cluster

In June of 2012, the US CDC initiated a multi-state outbreak investigation of *S*. Enteritidis associated with ground beef distributed principally in the northeast US (http://www.cdc.gov/salmonella/enteritidis-07-12/map.html). Shortly after the

Table 3.1 Samples linked to the LTCF outbreak by epidemiology and by WGS in the case study 1 cohort

Isolation date	County	Epidemiologically linked	Clustered by WGS
8/8/10	Nassau	–	–
8/10/10	Cattaraugus	–	–
8/16/10	Suffolk	–	–
8/22/10	Out-Of-State	–	–
8/26/10	Rockland	–	–
9/10/10	Putnam	–	+
9/10/10	Putnam	–	+
9/11/10	Putnam	–	+
9/12/10	Greenwich CT	+	+
9/12/10	Westchester	+	+
9/12/10	Westchester	+	+
9/13/10	Washington	–	+
9/13/10	Westchester	+	+
9/13/10	Westchester	+	+
9/13/10	Erie	–	–
9/14/10	Erie	–	–
9/15/10	Orange	–	–
9/16/10	Greenwich CT	+	+
9/16/10	Westchester	–	–
9/17/10	unknown	+	+
9/20/10	Westchester	–	+
9/22/10	Putnam	–	+
9/28/10	Putnam	–	+
10/4/10	Westchester	–	–
10/8/10	Putnam	–	+
10/27/10	Westchester	–	–
10/29/10	Nassau	–	+
2/1/11	Onondaga	–	–
2/21/11	Westchester	–	–
7/13/11	Rockland	–	–
7/22/11	Yates	–	–
9/6/11	Erie	–	–
10/5/11	Suffolk	–	–
10/9/11	Madison	–	–
10/22/11	Onondaga	–	–

+ Linked by epidemiology or clustered by WGS
– Putative sporadic isolate
All sample are PFGE pattern JEGX01.0004

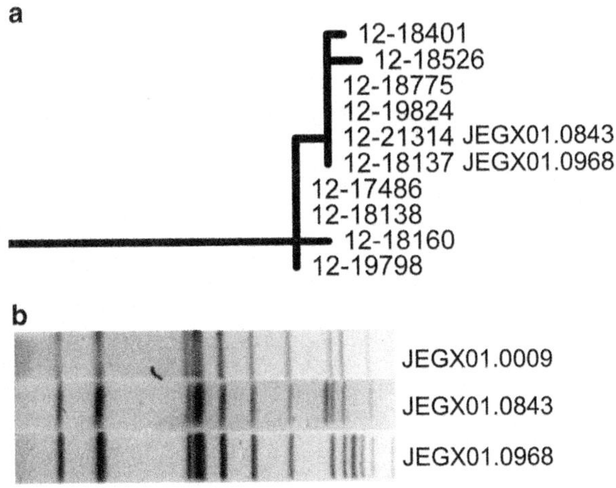

Fig. 3.2 Analysis of SE isolates associated with an outbreak from ground beef. (**a**) WGS SNP based phylogenetic analysis showing the clade harboring all of the outbreak isolates sequenced at the Wadsworth Center. There is 0–3 SNPs diversity in this clade. All isolates in the clade are JEGX01.0009 except as indicated. (**b**) PFGE of JEGX01.0009, JEGX01.0843 and JEGX01.0968 isolates from the clade. In the WGS analysis JEGX01.0843 and JEGX01.0968 are zero SNPS distant from two JEGX01.0009 isolates

outbreak began, 10 of the 20 *S.* Enteritidis isolates recently received at WCNYSDOH were sequenced [27]. Eight isolates were PFGE pattern JEGX01.0009, one was JEGX01.0843 and another isolate was JEGX01.0968. One of the isolates was isolated from the suspect hamburger (Fig. 3.2). The average SNP diversity within this outbreak was 1.2 SNPs (range of 0–3 SNPs). Interestingly, the SNP profile for the JEGX01.0843 and JEGX01.0968 (red arrows on figure) isolates were the same as two of the pattern JEGX01.0009 isolates, suggesting all four isolates had a common genomic backbone. *De novo* assembly of genomes of the JEGX01.0843 and JEGX01.0968 isolates revealed one or more large plasmids in each that were absent from all pattern JEGX01.0009 isolates in this clade. Thus, genomically defined outbreak clusters can harbor multiple PFGE subtypes. Furthermore this situation occurs frequently. In an ongoing study analyzing WGS *S.* Enteritidis, about 20 % of all WGS clusters (defined as well supported clades with six or less SNP diversity) have mixed PFGE types.

Case Study 3: a Real-Time Analysis of a Cluster Associated with Bean Sprouts

In the fall of 2014, the US CDC in collaboration with the US FDA, identified a US multi-state outbreak of *S.* Enteritidis. PFGE pattern JEGX01.0001, associated with bean sprouts (http://www.cdc.gov/salmonella/enteritidis-11-14/index.

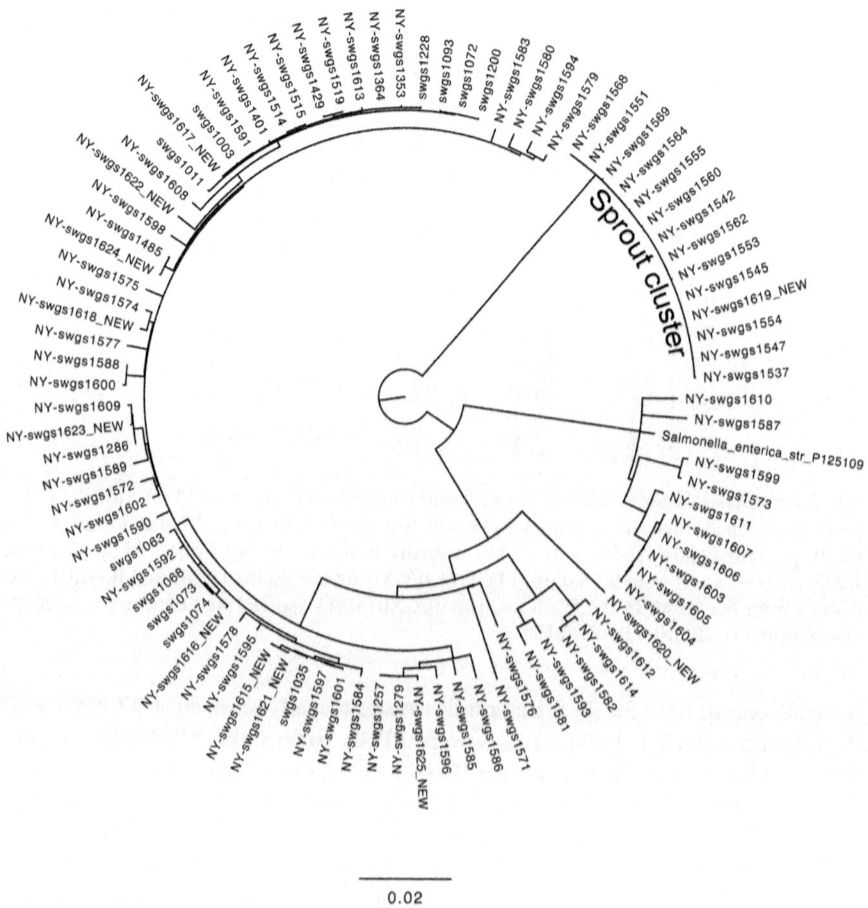

Fig. 3.3 WGS SNP based phylogenetic analysis dendrogram illustrating the tight clustering of isolates (0–3 SNPs diversity) associated with the bean sprout outbreak

html). This outbreak sickened 115 people in 12 US states, 22 of whom resided in NY. WCNYSDOH sequenced 14 patient isolates in real-time using the in house protocol (described above). The mean turn-around time from extraction of DNA to cluster analysis was 9.8 days (range of 8–15 days). All isolates fell in a well-supported clade with a mean SNP diversity of <1.0 SNP (0–3 SNPs range) (Fig. 3.3). Interestingly the last isolate we received (NY-swgs1619) was collected in mid-February, approximately 3 weeks after the outbreak appeared to be over.

For all three case studies, SNP diversity for an epidemiologically well-defined outbreak from a single source was remarkably low (range 0–3 with a mean of around 1 SNP). Similar results were obtained for seven other historical outbreaks from cases in MN with strong epidemiological data (Jones et al. unpublished).

Current State of Whole Generation Sequencing at Public Health Laboratories

In December 2014, the Association of Public Health Laboratories completed a survey of PHLs to understand the use of WGS in PHLs and concerns PHLs have with this new technology (http://www.aphl.org/AboutAPHL/publications/Documents/ID_NGSSurveyReport_52015.pdf#search=next%20generation%20survey%20results). The survey showed that 21 US state public health laboratories had WGS instrumentation and nine additional US state public health laboratories will purchase an instrument within the next 12 months. No local public health laboratories had a WGS instrument but three were planning on purchasing an instrument in the next 12 months. Reasons cited for not purchasing an instrument include: lack of funding, wanting applications to be more fully developed before purchasing, the expense of the instrument and insufficient staff to add new methods. The survey showed that PHLs are struggling to identify applications other than foodborne disease surveillance for WGS. Additional concerns from PHLs include data storage and transmission, potential privacy issues with sharing of metadata, data analysis and training.

Challenges to US PulseNet Laboratories of Adopting WGS

Cost and Efficiency

There are significant challenges for US PulseNet laboratories as the technology to perform surveillance evolves from PFGE to WGS. Cost is one of the principal concerns. New equipment and the associated reagents are expensive compared to traditional methods. Most PHLs do not have the funding in their budgets to purchase this equipment without help from another agency. Reagent costs are also quite significant. It is likely, but is not assured, that cost will decrease over time as technology evolves. Currently, it costs approximately 100 USD for reagents to perform WGS on a single *Salmonella* isolate. This price assumes optimal multiplexing of WGS (on a MiSeq sequence platform) balancing cost against number of sequencing reads needed to obtain the required depth and coverage of the genome. It does not include labor costs.

Some PHLs from less populated US states will have challenges achieving efficiency with WGS (see discussion above in WCNYSDOH workflow). Some US states only receive a few 100 isolates of US PulseNet organisms in a year. This creates a dilemma of performing smaller batches of WGS, which will increase the costs significantly, or delaying WGS until a large batch is available which would decrease timeliness of the testing. Delaying testing will make it more difficult to identify outbreaks. There are a few approaches that can be taken to increase efficiency while ensuring a reasonable turnaround time. One approach is to perform

WGS on non-PulseNet pathogens. *Mycobacterium tuberculosis*, viruses, and invasive bacteria are a few options. Another possibility to improve WGS efficiency is outsourcing; smaller PHLs could partner and send bacteria to larger PHLs to perform sequencing. This approach would not require purchasing of expensive WGS equipment and infrastructure and would not require advanced WGS training for lab staff at smaller PHLs. However, this approach would delay time to results and would prevent smaller PHLs from developing new technology infrastructure which may be detrimental as institutions in general adopt genomic technologies. It is likely that both approaches will be necessary in some US states.

Bioinformatic Analysis of Data

Additional challenges for PHLs as they adopt WGS technology is bioinformatic analysis, storage, and sharing of data. Although a variety of approaches for analysis of WGS data are well understood, a single approach that best serves a surveillance need has not been determined. Furthermore, it is likely that different organisms will benefit from using different analytical methods. Current WGS analysis requires a high level of expertise in bioinformatics as well as specialized software and infrastructure not available in most PHLs (see an exception in NYS in workflow section). Software to analyze WGS sequences that does not require a high level of bioinformatics is a necessity if WGS is going to be adopted as the subtyping method by PulseNet laboratories. Storage of data will likely occur at NCBI reducing the need for onsite storage capacity. Data transmission capacity must exist and has been a stumbling block for some PHLs implementation of WGS. Lastly, it is unlikely a single means for interpreting WGS will be arrived at soon. For now and into the near future, refining the interpretation of the output of these pipelines will be an iterative process (see below) as more samples and different organisms are analyzed prospectively.

As bacterial subtyping technology changes from PFGE to WGS, it is going to be critical to understand new subtyping techniques in an epidemiological context. The major goal of subtyping of foodborne pathogens is to identify cases that may have become ill from a common source. PFGE is the traditional subtyping method for PulseNet. Through years of investigation, public health laboratorians were able to interpret PFGE patterns to understand when investigation of a common source cluster was necessary [41–43]. Laboratorians were able to infer likely cases associations by the similarities of enteric bacteria isolated from the cases. While these interpretation guidelines may have not always created the correct answer, at least the rules that laboratories utilized were understood. With the advent of WGS, new interpretation guidelines will need to be developed. Question such as how many genetic differences (SNPs or alleles) expected to be seen among isolates during an outbreak, the amount of genetic differences one would see between outbreak and non-related isolates of the same species and serotype will be vital to understanding future technologies. These important questions can only be answered using isolates linked to well-characterized epidemiological data.

One strategy is to perform retrospective studies that include isolates from an outbreak (preferably several outbreaks), isolates that are known to not be related and isolates from an ill individual over time. These studies have been performed for *S.* Enteritidis ([27, 29], and case studies). Information obtained from these studies can be used to interpret WGS data, however it is likely that there will be different interpretation criteria for different species so these studies will need to be performed on multiple species and serotypes.

Once retrospective studies are performed to understand how to effectively interpret WGS data for outbreak investigations, prospective data analysis is needed on an on-going basis to iteratively validate outbreak cluster criteria. To achieve this requires good case exposure data. High quality epidemiological exposure data is not obtained in all jurisdictions but will be critical for evaluating new subtyping methods. Performing WGS without a standardized and tested strategy to interpret data will lead to poor harmonization of data across jurisdictions and a less effective system for surveillance.

Prioritization of Additional Clusters

WGS based surveillance vastly improves resolution and the granularity of cluster data. For instance, during surveillance of *S.* Enteritidis in NY state using WGS over a period of a year and a half, 40 genomic clusters were resolved with six SNPs or less diversity in the endemic JEGX01.0004 PFGE pattern. In total, 84 genomic clusters were identified over this time as compared to five non-endemic clusters detected by PFGE. In principle the improved resolution should aid epidemiologists in traceback investigations. While this improved resolution is welcome in a limited resource environment where it is impossible to follow up on all these leads, it also creates challenges. How to prioritize these clusters to allow efficient use of resources is an outstanding problem.

The laboratory could prioritize clusters for reporting to epidemiologists based on variables such as SNP diversity (how many SNPs between isolates to define a cluster), the number of isolates in a cluster (the more the better), length of time between acquiring isolates within the cluster (the shorter the better). This approach of identifying factors to prioritize cluster investigation has been performed successfully for *E. coli* 0157 [44] and *Salmonella* [45]. In a retrospective analysis of PFGE cluster characteristics that predicted successful epidemiological outcomes, Round's et al. [45] showed that clusters with four or more samples are more likely to result in identification of an outbreak source. In addition those in which three samples were received within a week were also more likely to be resolved. It seems reasonable to expect that similar criteria will need to be established for WGS data in order to manage the increased number of clusters detected and to establish priorities for their follow-up. Optimally, additional resources would be incorporated into foodborne epidemiology so all clusters could be investigated as soon as they are identified.

The WCNYSDOH analyzed the impact of changing the number of SNPs used to define a cluster (SNP diversity) in a phylogenetic dataset from *S.* Enteritidis

Table 3.2 Number of clusters detected as SNP diversity (number of SNPs between isolates in a cluster) is changed in a dataset from *S.* Enteritidis collected in real-time over a one and a half year period

SNP diversity	Number of clusters	Percent of clusters compared to SNP diversity of 5
0	38	45
1	63	76
2	76	94
3	83	101
4	85	105
5	84	100
6	83	101
7	81	102
8	82	99
9	81	101
10	84	96

collected in real-time over a period of 18 months. Lowering the SNP diversity threshold from 5 to 0 SNPs resulted in detecting about 45 % fewer cluster being detected (Table 3.2). But this effect was not linear and once SNP diversity was raised to two SNPs there was little change in the number of clusters detected. Because we know from retrospective studies that outbreaks can have a SNP diversity of up to three SNPs, lowering our cutoff below this level will likely lead to missing outbreak cases. Hence reducing SNP diversity may not be a useful means to prioritize clusters.

QA/QC

WGS technologies and data analysis and interpretation are likely to evolve significantly in the next few years. Improved turn-around time, reduced costs, and increased accuracy can be expected. However, this creates a challenge to surveillance systems based on the technology that is evolving rather than a stable technology such as PFGE.

To achieve continuity within the public health network, PHLs are required to maintain quality standards for all laboratory testing to ensure the accuracy, reliability and timeliness of patient test results regardless of where the test was performed. Most PHLs use Clinical Laboratory Improvement Amendments (CLIA) or College of American Pathologists (CAP) standards to evaluate tests quality. Expected quality measurements will need to be established for WGS subtyping methods for surveillance so that inter-laboratory comparison of sequences can be analyzed. At this time such standards do not exist. Further complicating the development of standards is that it is expected the WGS methods, reagents, databases and equipment will change significantly in the upcoming years. In view of this it is essential that physical as well as analytical standards be developed and validated so that future performance can be interpreted in relation to past performance. National Institute of Standards and Technology (NIST), NCBI, US CDC, and US FDA in collaboration with state institutions are starting to address this issue.

Similarly, it will be a challenge for PHLs to stay current on advancing methods and technologies while still maintaining performance standards. It will be important for PHLs to thoughtfully consider how they will respond to technology advances to stay current and to ensure the quality of their testing.

The Promise of Uniform Workflows and the Reality

Surveillance by WGS holds the promise of a simplified workflow. With the exception of DNA extraction, which varies based on cell wall properties, and the number of templates sequenced (depends on genome size and desired sequencing depth) protocols for a given sequencing platform can be highly uniform for all bacteria assayed. In contrast, PFGE requires different enzymes for each organism and MLST and MLVA require specific primer sets for each organism precluding processing multiple species in a single workflow. Once WGS is completed, data transmission, analysis, reporting may also be completed in a standardized manner simplifying these processes as well. Thus once surveillance has transitioned to WGS uniform workflows in all aspects of the process should produce cost and time savings, increased capacity, and as described above improved resolution of outbreaks and outcomes.

To achieve the promise of uniform workflows, best practices need to be established. Best practices can only be agreed upon once standard quality metrics are established (see above). At this time federal and state entities have established standards in an ad hoc manner. This results in extra work for implementation of WGS in the public health laboratory that must accommodate varying protocols and standards depending on whom they are serving. For instance, at the WCNYSDOH, transmission and analysis of sequence data performed on patient specimens for the US CDC or for internal projects are handled differently (see workflow description above). For GenomeTrakr studies, yet another workflow is implemented. This results in some redundancy and inefficient use of resources. However, this somewhat chaotic state of affairs can be tolerated for the time being with the expectation such experimentation will inform the establishment of future best practices for uniform workflows.

The genomic era of public health science is here. Abundant data now exists to demonstrate that the technology will improve subtyping for surveillance of clonal organisms. Furthermore, increased efficiencies are expected as the same datasets can be parsed to identify serotype, virulence markers, and antibiotic resistance mechanisms. Yet a great deal of uncertainty still surrounds the endeavor and causes concern among practitioners. As we move forward we must implement standards and determine best practices. Without the assurance of these beacons of the past, the way forward is unclear and rapid progress to achieve the promise of the microbial genomic era is jeopardized.

References

1. Scallan E, Mahon BE, Hoekstra RM, Griffin PM. Estimates of illnesses, hospitalizations and deaths caused by major bacterial enteric pathogens in young children in the United States. Pediatr Infect Dis J. 2013;32(3):217–21. doi:10.1097/INF.0b013e31827ca763.
2. Snitkin ES, Zelazny AM, Thomas PJ, Stock F, NCSP Comparative Sequencing Program Group, Henderson DK, et al. Tracking a hospital outbreak of carbapenem-resistant *Klebsiella pneumoniae* with whole-genome sequencing. Sci Transl Med. 2012;4(148), 148ra16. doi:10.1126/scitranslmed.3004129.
3. Stucki D, Ballif M, Bodmer T, Coscolla M, Maurer AM, Droz S, et al. Tracking a tuberculosis outbreak over 21 years: strain-specific single nucleotide polymorphism-typing combined with targeted whole genome sequencing. J Infect Dis. 2014;211(8):1306–16. doi:10.1093/infdis/jiu601.
4. Chinthapalli K. DNA sequencing helped to limit spread of MRSA in a neonatal unit. BMJ. 2012;345, e7746. doi:10.1136/bmj.e7746.
5. Tellez-Sosa J, Rodriguez MH, Gomez-Barreto RE, Valdovinos-Torres H, Hidalgo AC, Cruz-Hervert P, et al. Using high-throughput sequencing to leverage surveillance of genetic diversity and oseltamivir resistance: a pilot study during the 2009 influenza A(H1N1) pandemic. PLoS One. 2013;8(7), e67010. doi:10.1371/journal.pone.0067010.
6. Bell A, Lewandowski K, Myers R, Wooldridge D, Aarons E, Simpson A, et al. Genome sequence analysis of Ebola virus in clinical samples from three British healthcare workers, August 2014 to March 2015. Euro Surveill. 2015;20(20), pii=21131.
7. Volz E, Pond S. Phylodynamic analysis of Ebola virus in the 2014 Sierra Leone epidemic. PLoS Curr. 2014;6. doi:10.1371/currents.outbreaks.6f7025f1271821d4c815385b08f5f80e.
8. Rodrigue DC, Cameron DN, Puhr ND, Brenner FW, St Louis ME, Wachsmuth IK, et al. Comparison of plasmid profiles, phage types, and antimicrobial resistance patterns of *Salmonella* Enteritidis isolates in the United States. J Clin Microbiol. 1992;30(4):854–7.
9. Boxrud D, Pederson-Gulrud K, Wotton J, Medus C, Lyszkowicz E, Besser J, et al. Comparison of multiple-locus variable-number tandem repeat analysis, pulsed-field gel electrophoresis, and phage typing for subtype analysis of *Salmonella enterica* serotype Enteritidis. J Clin Microbiol. 2007;45(2):536–43. doi:10.1128/JCM.01595-06.
10. Clark CG, Kruk TM, Bryden L, Hirvi Y, Ahmed R, Rodgers FG. Subtyping of *Salmonella enterica* serotype Enteritidis strains by manual and automated PstI-SphI ribotyping. J Clin Microbiol. 2003;41(1):27–33.
11. Ridley AM, Threlfall EJ, Rowe B. Genotypic characterization of *Salmonella Enteritidis* phage types by plasmid analysis, ribotyping, and pulsed-field gel electrophoresis. J Clin Microbiol. 1998;36(8):2314–21.
12. Liebana E, Clouting C, Garcia-Migura L, Clifton-Hadley FA, Lindsay E, Threlfall EJ, et al. Multiple genetic typing of *Salmonella* Enteritidis phage-types 4, 6, 7, 8 and 13a isolates from animals and humans in the UK. Vet Microbiol. 2004;100(3–4):189–95. doi:10.1016/j.vetmic.2004.01.020.
13. Desai M, Threlfall EJ, Stanley J. Fluorescent amplified-fragment length polymorphism subtyping of the Salmonella enterica serovar Enteritidis phage type 4 clone complex. J Clin Microbiol. 2001;39(1):201–6. doi:10.1128/JCM.39.1.201-206.2001.
14. Scott F, Threlfall J, Stanley J, Arnold C. Fluorescent amplified fragment length polymorphism genotyping of *Salmonella* Enteritidis: a method suitable for rapid outbreak recognition. Clin Microbiol Infect. 2001;7(9):479–85.
15. Boxrud D, Monson T, Stiles T, Besser J. The role, challenges, and support of pulse net laboratories in detecting foodborne disease outbreaks. Public Health Rep. 2010;125 Suppl 2:57–62.
16. Swaminathan B, Barrett TJ, Hunter SB, Tauxe RV, Force CDCPT. PulseNet: the molecular subtyping network for foodborne bacterial disease surveillance, United States. Emerg Infect Dis. 2001;7(3):382–9. doi:10.3201/eid0703.010303.

17. McCollum JT, Cronquist AB, Silk BJ, Jackson KA, O'Connor KA, Cosgrove S, et al. Multistate outbreak of listeriosis associated with cantaloupe. N Engl J Med. 2013;369(10):944–53. doi:10.1056/NEJMoa1215837.
18. Cavallaro E, Date K, Medus C, Meyer S, Miller B, Kim C, et al. Salmonella typhimurium infections associated with peanut products. N Engl J Med. 2011;365(7):601–10. doi:10.1056/NEJMoa1011208.
19. From the Centers for Disease Control and Prevention. Multistate outbreak of *Escherichia coli* O157:H7 infections associated with eating ground beef—United States, June–July 2002. JAMA. 2002;288(6):690–1.
20. Quick J, Ashton P, Calus S, Chatt C, Gossain S, Hawker J, et al. Rapid draft sequencing and real-time nanopore sequencing in a hospital outbreak of *Salmonella*. Genome Biol. 2015;16(1):114. doi:10.1186/s13059-015-0677-2.
21. Bakker HC, Switt AI, Cummings CA, Hoelzer K, Degoricija L, Rodriguez-Rivera LD, et al. A whole-genome single nucleotide polymorphism-based approach to trace and identify outbreaks linked to a common *Salmonella enterica* subsp. enterica serovar Montevideo pulsed-field gel electrophoresis type. Appl Environ Microbiol. 2011;77(24):8648–55. doi:10.1128/AEM.06538-11.
22. Lienau EK, Blazar JM, Wang C, Brown EW, Stones R, Musser S, et al. Phylogenomic analysis identifies gene gains that define *Salmonella enterica* subspecies I. PLoS One. 2013;8(10), e76821. doi:10.1371/journal.pone.0076821.
23. Lienau EK, Strain E, Wang C, Zheng J, Ottesen AR, Keys CE, et al. Identification of a salmonellosis outbreak by means of molecular sequencing. N Engl J Med. 2011;364(10):981–2. doi:10.1056/NEJMc1100443.
24. Hoffmann M, Luo Y, Monday SR, Gonzalez-Escalona N, Ottesen AR, Muruvanda T, et al. Tracing origins of the *Salmonella* Bareilly strain causing a food-borne outbreak in the United States. J Infect Dis. 2015;213(4):502–8. doi:10.1093/infdis/jiv297.
25. Fey PD, Iwen PC, Zentz EB, Briska AM, Henkhaus JK, Bryant KA, et al. Assessment of whole-genome mapping in a well-defined outbreak of *Salmonella enterica* serotype Saintpaul. J Clin Microbiol. 2012;50(9):3063–5. doi:10.1128/JCM.01320-12.
26. Angelo KM, Chu A, Anand M, Nguyen TA, Bottichio L, Wise M, et al. Outbreak of *Salmonella* Newport infections linked to cucumbers—United States, 2014. MMWR. 2015;64(6):144–7.
27. den Bakker HC, Allard MW, Bopp D, Brown EW, Fontana J, Iqbal Z, et al. Rapid whole-genome sequencing for surveillance of *Salmonella enterica* serovar Enteritidis. Emerg Infect Dis. 2014;20(8):1306–14. doi:10.3201/eid2008.131399.
28. Joensen KG, Scheutz F, Lund O, Hasman H, Kaas RS, Nielsen EM, et al. Real-time whole-genome sequencing for routine typing, surveillance, and outbreak detection of verotoxigenic *Escherichia coli*. J Clin Microbiol. 2014;52(5):1501–10. doi:10.1128/JCM.03617-13.
29. Taylor AJ, Lappi V, Wolfgang WJ, Lapierre P, Palumbo MJ, Medus C, Boxrud D. Characterization of foodborne outbreaks of *Salmonella enterica* serovar Enteritidis with whole genome sequencing SNP-based analysis for surveillance and outbreak detection. J Clin Microbiol. 2015;53(10):3334–40.
30. Gerner-Smidt P, Kincaid J, Kubota K, Hise K, Hunter SB, Fair MA, et al. Molecular surveillance of Shiga toxigenic *Escherichia coli* O157 by PulseNet USA. J Food Prot. 2005;68(9):1926–31.
31. van Belkum A, Tassios PT, Dijkshoorn L, Haeggman S, Cookson B, Fry NK, et al. Guidelines for the validation and application of typing methods for use in bacterial epidemiology. Clin Microbiol Infect. 2007;13 Suppl 3:1–46. doi:10.1111/j.1469-0691.2007.01786.x.
32. Deng X, Shariat N, Driebe EM, Roe CC, Tolar B, Trees E, et al. Comparative analysis of subtyping methods against a whole-genome-sequencing standard for *Salmonella enterica* serotype Enteritidis. J Clin Microbiol. 2015;53(1):212–8. doi:10.1128/JCM.02332-14.
33. Dallman TJ, Byrne L, Ashton PM, Cowley LA, Perry NT, Adak G, et al. Whole genome sequencing for national surveillance of Shiga toxin producing *Escherichia coli* O157. Clin Infect Dis. 2015;61(3):305–12. doi:10.1093/cid/civ318.
34. Jenkins C, Dallman TJ, Launders N, Willis C, Byrne L, Jorgensen F, et al. Public health investigation of two outbreaks of Shiga toxin-producing *Escherichia coli* O157 associated with

consumption of watercress. Appl Environ Microbiol. 2015;81(12):3946–52. doi:10.1128/AEM.04188-14.

35. Underwood AP, Dallman T, Thomson NR, Williams M, Harker K, Perry N, et al. Public health value of next-generation DNA sequencing of enterohemorrhagic *Escherichia coli* isolates from an outbreak. J Clin Microbiol. 2013;51(1):232–7. doi:10.1128/JCM.01696-12.
36. Byrne L, Fisher I, Peters T, Mather A, Thomson N, Rosner B, et al. A multi-country outbreak of *Salmonella* Newport gastroenteritis in Europe associated with watermelon from Brazil, confirmed by whole genome sequencing: October 2011 to January 2012. Euro Surveill. 2014;19(31):6–13.
37. Ashton PM, Peters T, Ameh L, McAleer R, Petrie S, Nair S, et al. Whole genome sequencing for the retrospective investigation of an outbreak of Salmonella Typhimurium DT 8. PLoS Curr. 2015;7. doi:10.1371/currents.outbreaks.2c05a47d292f376afc5a6fcdd8a7a3b6.
38. Dallman TJ, Byrne L, Launders N, Glen K, Grant KA, Jenkins C. The utility and public health implications of PCR and whole genome sequencing for the detection and investigation of an outbreak of Shiga toxin-producing *Escherichia coli* serogroup O26:H11. Epidemiol Infect. 2015;143(8):1672–80. doi:10.1017/S0950268814002696.
39. Gilmour MW, Graham M, Van Domselaar G, Tyler S, Kent H, Trout-Yakel KM, et al. High-throughput genome sequencing of two Listeria monocytogenes clinical isolates during a large foodborne outbreak. BMC Genomics. 2010;11:120. doi:10.1186/1471-2164-11-120.
40. Silk BJMB, Griffin PM, Gould LH, Tauxe RV, Crim SM, Jackson KA, Gerner-Smidt P, Herman KM, Henao OL. Listeri illnesses, deaths, and outbreaks—United States, 2009–2011. Morb Mortal Wkly Rep. 2013;62(22):448–52.
41. Barrett TJ, Gerner-Smidt P, Swaminathan B. Interpretation of pulsed-field gel electrophoresis patterns in foodborne disease investigations and surveillance. Foodborne Pathog Dis. 2006;3(1):20–31. doi:10.1089/fpd.2006.3.20.
42. Tenover FC, Arbeit RD, Goering RV, Mickelsen PA, Murray BE, Persing DH, et al. Interpreting chromosomal DNA restriction patterns produced by pulsed-field gel electrophoresis: criteria for bacterial strain typing. J Clin Microbiol. 1995;33(9):2233–9.
43. Goering RV, Kock R, Grundmann H, Werner G, Friedrich AW, ESCMID Study Group for Epidemiological Markers (ESGEM). From theory to practice: molecular strain typing for the clinical and public health setting. Euro Surveill. 2013;18(4):20383.
44. Rounds JM, Boxrud DJ, Jawahir SL, Smith KE. Dynamics of *Escherichia coli* O157:H7 outbreak detection and investigation, Minnesota 2000–2008. Epidemiol Infect. 2012;140(8):1430–8. doi:10.1017/S0950268811002330.
45. Rounds JM, Hedberg CW, Meyer S, Boxrud DJ, Smith KE. *Salmonella enterica* pulsed-field gel electrophoresis clusters, Minnesota, USA, 2001–2007. Emerg Infect Dis. 2010;16(11):1678–85. doi:10.3201/eid1611.100368.

Chapter 4
Bioinformatics Aspects of Foodborne Pathogen Research

Henk C. den Bakker, Laura K. Strawn, and Xiangyu Deng

Introduction

In the early days of the genome revolution Luscombe et al. proposed to define bioinformatics as '(the field) conceptualizing biology in terms of macromolecules (in the sense of physical-chemistry) and then applying "informatics" techniques (derived from disciplines such as applied mathematics, computer science, and statistics) to understand and organize the information associated with these molecules, on a large-scale' [1]. Translated to the present day, bioinformatics is a field that involves algorithm-, pipeline- and software development, analysis, transfer and storage/database development of omics data. Most of these aspects of bioinformatics are discussed in other chapters of this book; Good examples of web-based pipelines can be found in Chapter 5, while examples of the bioinformatics aspects of RNA sequencing are given in Chapter 10.

In this chapter we will focus on several general aspects of bioinformatics approaches and how they are used to probe different aspects of the biology and epidemiology of foodborne pathogens.

H.C. den Bakker (✉)
Department of Animal and Food Sciences, Texas Tech University, Lubbock, TX, USA
e-mail: henk.c.den-bakker@ttu.edu

L.K. Strawn
Department of Food and Technology, Eastern Shore Agricultural Research and Extension Center, Virginia Tech, Painter, VA, USA

X. Deng
Center for Food Safety, University of Georgia, Griffin, GA, USA

X. Deng et al. (eds.), *Applied Genomics of Foodborne Pathogens*, Food Microbiology and Food Safety, DOI 10.1007/978-3-319-43751-4_4

Sequencing Technologies

The last decade has seen the emergence of several Next Generation Sequencing (NGS) technologies, also referred to as high-throughput sequencing technologies, as opposed to the until then low throughput, but dominant, automated Sanger technology [2]. The most prominent improvement of NGS technologies on automated Sanger technology is that they produce massive amounts of data at greatly reduced per base-pair sequencing cost, making it possible to sequence complete microbial genomes at a price comparable to traditional subtyping methods such as Pulsed Field Gel Electrophoresis (PFGE) or multi-locus sequence typing (MLST). While several technologies have been commercially available (see [2] for a review), three technologies currently dominate the field of microbial genomics; (1) Illumina, (2) IonTorrent, and (3) Pacific Biosciences. Both Illumina and Ion Torrent can be classified as short read sequence technologies, producing large numbers of short (up to 300 bp for Illumina and 400 bp for Ion Torrent) reads. The Ion Torrent technology relies on so-called 'semiconductor' sequencing, while the Illumina platform relies on a 'sequencing by synthesis' based technology, using reversible (fluorescent) terminators. The Ion Torrent is prone to homopolymeric tracts length related errors [3], and while reads generated on any of the Illumina sequencers are characterized by a high accuracy, they are not free from errors, and may display more subtle sequence motif related errors [4, 5]. Pacific Biosciences sequencing relies on a single molecule real-time (SMRT) sequencing technology. This sequencing platform has the possibility to produce long reads (>10 kbp) with relatively low per read accuracy, however, consensus accuracy is high [6]. In addition to base composition of sequences, base modifications (e.g., methylation) can be inferred [7] from data generated with SMRT sequencing.

Bioinformatics Approaches

Raw sequencing reads need to be properly processed before biological information can be extracted and interpreted. Two categories of bioinformatics methods are often used to analyze microbial sequencing data, read mapping approaches and de novo approaches.

Read Mapping Based Approaches

Read mapping based approaches rely on the alignment (mapping) of large numbers (typically in the hundreds of thousands) of raw sequence reads to a (preferably completely sequenced) reference genome. Bowtie 2 [8] and BWA [9] are the most commonly used read mapping algorithms for short read data, such as those produced by the Illumina platforms. Both algorithms rely on a Burrow-Wheeler transform to index the reference genome, followed by alignment of the sequence reads using a relaxed, quality-aware algorithm [8, 9]. Due to the relative speed and low memory requirements, read mapping approaches are commonly used in small genomic variant (single nucleotide polymorphisms, small insertions or deletions) detection pipelines [10, 11]. Another important application relying on reference mapping is RNA seq as discussed in Chapter 10.

De Novo Approaches

Some major disadvantages of read mapping based approaches is the reliance on a good reference genome; appropriate finished reference genome sequences (i.e., genome sequences of strains closely related to the subtype being researched) are not always available, genomic regions present in the query strain but not in the reference strain will not be included in the analyses and regions of high divergence may make the mapping process difficult and downstream analyses less reliable [12]. Several approaches have been developed to infer genomic variants de novo which require minimal or no assembly of the original reads. An example of one of these approaches is kSNP [13, 14], an k-mer based approach, or Cortex_var, an approach which relies on the construction of de Bruijn graphs to infer genomic variants [15, 16].

De Novo Assembly and the Basics of Comparative Genomics

De novo assembly consists of the (partial) reconstruction of a genome out of raw reads. Currently, most modern assemblers rely on one of two algorithmic principles; (1) overlap, layout and consensus (OLC) approaches and approaches relying on an assembly graph, in most cases a De Bruijn graph. OLC-based methods have traditionally been used for Sanger sequencing datasets and are currently used in assembly software such as Celera [17], EDENA [18] and MIRA [19]. OLC-based assemblers are currently making a resurgence with the emergence of ultra-long (i.e. several kilo-basepairs or more) sequence read technologies, such as Pacific BioSciences and Oxford Nanopore, which can greatly facilitate de novo genome assembly with or without the polishing of high quality short reads to correct the still error-prone long reads The most commonly used software for de novo assembly is De Bruijn graph based. Reads are broken up in smaller kmers, and these kmers are used to construct an assembly graph from which the final genome assembly will be deduced. In a theoretical situation, this graph would form a perfect circle representing a bacterial chromosome or plasmid; however, in reality due to the existence of repeat regions and sequencing errors this graph becomes more complex, displaying 'bubbles' due to sequence errors in the middle of reads, cycles due to repetitive sequence and spurs, caused by sequence errors at the end of a read. A critical factor in the use of de Bruijn based assemblers is the choice of the length of the k-mer, being of importance for the length and accuracy of the final contigs in the assembled genome. While this was particularly critical for early De Bruyn assemblers, such as Velvet [20], this issue has been largely overcome by the use of multiple k-mers in the approaches used in next generation De Bruijn assemblers such as SPAdes [21].

Comparative Genomics

Genome annotation is the process of identifying and labeling relevant genomic features on a genomic sequence [22, 23]. While the manual annotation of a bacterial genome was a long and labor-intensive process in the early days of genome sequencing [24–26], automated annotations can be generated within days or minutes with current tools such as RAST [27] and Prokka [23]. Most annotation pipelines use software such as Glimmer [28] or Prodigal [29] for initial model-based gene prediction. Coding sequences are then annotated by similarity to (preferably curated) databases and additional information of other features (such as rRNAs and tRNA) predicted by additional software tools can be added. A hierarchal approach of gene annotation is employed in Prokka [23], starting with annotation by sequence homology to a smaller trustworthy database, moving to annotation based on medium-sized but domain-specific databases, and finally to curated models of protein families. Genes that do not match any of these gene annotation models are then annotated as hypothetical proteins.

Automated genome annotations can be used for additional gene orthology based research, such as research focused on the pan-genome of a genus or species [30]. Traditionally this kind of analyses is based on orthology searches to determine the presence/absence of gene families in the genomes of related bacterial strains. Most orthology searches, such as those implemented in OrthoMCL [31] or the ITEP pipeline [32], employ the Markov Cluster Algorithm (MCL) [33], an algorithm that can be used to perform clustering of networks. This algorithm has been used for more than a decade now for the assignment of proteins into families based on precomputed sequence similarity information (i.e., BLAST or other similarity distances). Most pipelines rely on an initial all-against-all comparison with BLAST or BLASTP [34], making orthology analyses computationally intensive and therefore prohibitive when larger numbers of genomes are involved. Roary [35], employs CD-HIT [36] to rapidly cluster sequences before the BLAST step, thereby reducing the numbers of sequences that are used in the BLAST and MCL step, which makes it possible to perform analyses on hundreds of genomes with limited computational resources and with a limited amount of time. Results of gene orthology analyses are typically gene presence matrices and/or gene alignments, which can be used for further sequence based analyses (e.g., recombination and Darwinian selection analyses).

Phylogenetics

Phylogenetics studies the evolutionary history of groups of organisms, such as populations and species [37]. These evolutionary histories are inferred on the basis of homologous information, which translates mostly to aligned sequences (nucleotide or amino acid sequences) in bacterial genomes, but can also be based on binary patterns of gene presence or absence as inferred by the orthology analyses described in the previous section. Evolutionary relationships are generally depicted as phylogenetic trees or networks [38]. Methods of phylogenetic tree inference differ based on the optimality criterion used to determine the most plausible tree given the data. The most commonly used criteria are (1) parsimony, a criterion which favors the tree minimizing the evolutionary changes, (2) distance criteria, favoring trees minimizing evolutionary distances, such as total SNP differences or distances adjusted using evolutionary models, (3) the maximum likelihood criterion, which favors the tree maximizing the likelihood of observing the data given a model of molecular evolution and (4) Bayesian methods, which favor the trees with the highest posterior probability given the data and a model of molecular evolution (Felsenstein 2004). While generally different optimally criteria lead to similar phylogenetic trees when the phylogenetic signal of the dataset is strong, different optimality criteria suffer from specific artifacts under different scenarios. An example of such an artifact is long-branch attraction; an artifact which is observed in parsimony (Felsenstein 2004) and Bayesian phylogenetics [39] in which topologies that group long branches together are favored. Maximum likelihood methods seem most robust to these artifacts. While implementations of maximum likelihood used to be very computationally intensive, rapid algorithms ML algorithms have been introduced such as RAxML [40], Garli [41] and PhyML [42], making ML a feasible option for phylogenetic inference in bacterial genomics and outbreak surveillance.

A current common practice in ML and Bayesian phylogenetics based on bacterial whole genome data is the use of input matrices which consist entirely of variable sites from a whole genome alignment, whereas millions of homologous (mostly invariant) sites are excluded from the analysis. Sequence data without invariable sites are suffering from an acquisition or ascertainment bias [43] and most models of molecular evolution make the assumption that invariable sites are included in the analysis, and exclusion of invariable sites without proper correction for this bias can lead to branch length overestimation and biases in phylogeny inference [44]. RAxML [40] and BEAST [45] are examples of current software which have implemented the ability to use models that account for ascertainment bias. While ascertainment bias may not be a problem in phylogenetics when applied in outbreak investigations, where the prime goal is to identify clusters of closely related strains potentially involved in an outbreak, it may be more of an issue in studies trying to infer the evolutionary history of a group of organisms.

Homologous Recombination

Phylogenetic methods, with the exception of certain network approaches [38], operate under the assumption that the evolutionary history of a group of organisms can be represented as a non-reticulate, bifurcating tree. Homologous recombination, the replacement of an endogenous genomic region with a piece homologous, exogenous DNA, usually from a closely related organism, violates this principle, and could potentially distort phylogenetic inference, leading to biased estimates of branch lengths, artifactual signals of population expansion [46], false inference of positive selection [47, 48], and unreliable reconstruction of the tree topology [49]. Several bioinformatics tools (e.g., ClonalFrameML [50], Gubbins [51], BRATNextGen [52]) have been developed to detect and in most cases correct for the effect of homologous recombination in genome scale phylogenies. Interestingly, Hedge and Wilson [53] show that while homologous recombination badly distorts branch lengths in phylogenetic trees, the accuracy of the reconstruction of evolutionary relationships seems to be robust to the influence of homologous recombination.

Genomic Epidemiology of Foodborne Pathogens

While WGS and bioinformatics have been applied to studying foodborne pathogens in the early 2000s [24, 25, 54], it was the emerging practice of NGS technologies in foodborne pathogen surveillance and outbreak investigation that started to change the landscape of food safety and public health microbiology. The term "genomic epidemiology" has been increasingly used to describe the application of NGS in accessing, indexing and analyzing DNA sequence features of epidemiologic importance. Genomic elements that vary in rates of mutation in bacterial evolution supply ample targets for epidemiologic investigations at different temporal and geographical scales. Genetic determinants of certain phenotypes (e.g., serotype and antimicrobial resistance) provide the possibility of in silico prediction of clinically important phenotypic traits. Since whole genome sequences contain the entirety of genetic information of an organism, of which DNA markers of any sort become mere subsets, NGS promises a comprehensive platform for public health microbiology that provides a one-stop shop for various subtyping and characterization targets. For example, WGS can both help solve outbreaks by affording high discriminatory power in differentiating closely related isolates and assist in tracking epidemiologic trends by monitoring antibiotic/antimicrobial resistance. WGS also comes with backward compatibility and future extensibility to any existing and potential typing schemes as the complete set of DNA variations across entire genomes are made available for analysis.

In depth discussion of public health applications of NGS can be found in Chapters 1, 2, and 3 of this book and examples of bioinformatics tools for genomic epidemiology investigation will be provided in Chapter 5.

Routine and widespread application of WGS in foodborne pathogen surveillance requires methodological standards that can be practiced among laboratories. Rapid evolution of sequencing technologies and their analytical roots in bioinformatics make standardization a particular challenge. Comparative studies have been frequently performed to evaluate major and newly developed sequencers [3, 55]. While Illumina platforms have become the de facto standard for foodborne illness diagnosis and surveillance, new generation technologies are emerging such as the Oxford Nanopore based, thumb drive sized MinIon device whose utility in outbreak investigation has been recently demonstrated [56]. Compared with the evaluation of sequencing technologies where explicit quality attributes are available, benchmarking of bioinformatics tools for WGS data analysis can be more challenging. For general analyses including de novo genome assembly, read mapping and variants calling, multiple tools exist and oftentimes no single tool outperforms others with all data sets [57]. Further complexity can be generated when several software tools are combined into a workflow or pipeline—the workforce in public health laboratories that turns WGS data into epidemiologically relevant results such as phylogenetic trees. For example, the United States Centers for Disease Control and Prevention and the United States Food and Drug Administration each has individually developed a pipeline for SNP-based subtyping and phylogenetic analysis [10, 11].

A particular debate regarding the standard bioinformatics approach for WGS-based subtyping of foodborne pathogens has been focused on two methods, whole genome SNP typing (WGST) and whole genome multilocus sequencing typing (wgMLST). WGST has been the dominant method for phylogenetic analysis of WGS data. As stable phylogenetic markers [58], SNPs can be identified from whole genomes to allow robust evolutionary analysis and high resolution subtyping. However, WGST does not produce easily communicable nomenclatures. The set of polymorphic sites for subtyping, usually SNPs located in conserved parts of the genomes (i.e., core genome) to be analyzed, is subject to change when different groups of strains are investigated, making definitive naming of subtypes difficult. As standard nomenclature is important for coordinated surveillance, wgMLST has been recently proposed as a whole genome scale upgrade from the well-established multilocus sequencing typing (MLST) system [59]. A core genome (i.e., genes conserved among all strains of, typically, a bacterial species) version of wgMLST (core genome MLST or cgMLST) has been developed for *L. monocytogenes* [60]. In comparison with WGST, wgMLST allows definitive naming of sequence types by querying against a precompiled database of alleles. However, wgMLST may not be as discriminatory as WGST since intergenic regions with informative mutations are typically excluded from allele databases. Also, wgMLST can lead to simplification of allelic differences as multiple mutations can be collapsed into a single sequence type. Despite the shortcomings and the fact that allele databases for major foodborne pathogens are still being developed, it is anticipated that wgMLST will become the routine and primary method for WGS-powered pathogen surveillance at PulseNet (personal communication with Peter Gerner-Smidt).

Microbiome Research and Foodborne Pathogens

Next generation sequencing (NGS) driven microbiome research has revolutionized our understanding of microbial ecology and the role of the microbiome in human health. One of the big advantages of methods used in microbiome methods is that they represent so-called culture-free methods, i.e., these methods do not rely on selective media or other culturing steps, making it possible to study unculturable organisms and study otherwise culturable organisms without the bias introduced by the use of selective media. While culture free methods hold great promise, they have yet to be widely adopted in research on foodborne pathogens.

General Principles and Pipelines

Application of NGS in microbiome research generally relies on one of two strategies; (1) amplicon sequencing or (2) sequencing of total DNA of a microbial community, also known as metagenomics or "full shotgun metagenomics" [61]. Both approaches start with the extraction of total DNA of a sample of interest. Care should be taken to use an extraction protocol that efficiently extracts DNA from a wide taxonomic variety of microorganisms, without bias for certain taxonomic groups [62]. Amplicon sequencing starts with the PCR amplification, generally one of the hypervariable subregions of the small ribosomal subunit (16S) for bacteria and subsequent generation of sequencing libraries, while shotgun metagenomic libraries are made without an amplification step. Both approaches have their advantages and disadvantages; Amplicon sequencing requires smaller quantities of input DNA, and because a specific gene can be targeted (e.g., using generic primers targeting prokaryotic 16S regions), most of the sequence output will be derived from the microbiome, without inclusion of sequenced DNA of a host of DNA associated with a food source. Using PCR amplification of a specific target can however bias the obtained microbiome data to species that are more easily amplified with the primers used, even though generic primers are used. Another disadvantage is that the 16S target region is not informative enough to distinguish between taxa that may be of interest in food safety research. The small (<500 bp) 16S regions generally sequenced in Illumina experiments do not contain enough information to distinguish between genera in the Enterobacteriaceae, which contains pathogens such as *Salmonella enterica* and *Escherichia coli*. This can be potentially overcome in the near future if long read technologies (Pacific Biosciences SMRT sequences, Oxford Nanopore) are used; however, the use of a single gene may still be problematic in foodborne pathogen research. For instance, the divergence observed in the entire (~1500 bp) 16S rRNA region between pathogenic and non-pathogenic *Listeria* species (such as *L. monocytogenes* and *L. innocua*) is extremely low, making it difficult to distinguish them based on rRNA regions alone [63].

Popular pipelines used for 16S amplicon datasets are QIIME [64] and Mothur [65]. Both pipelines apply clustering algorithms (such as uClust or uSearch [66]) at specific levels of sequence divergence to assign sequence reads to individual Operational Taxonomic Units (OTUs). Alternatively reads can be mapped against a reference database such as Greengenes [67], RDP [68] or one of the curated databases of the National Center for Biotechnology Information. Dependent on the database used this approach can be applied to amplicon as well as full shotgun metagenomics and is applied in software such as Megan [69].

An alternative approach in full shotgun metagenomics is to assemble the reads into contigs, which can then be more accurately identified to species. While general purpose assemblers including SPAdes [21] can be used for this purpose, computationally efficient assemblers such as MEGAHIT [70] and metaSPAdes [71] are more suitable for this approach. Assemblies usually result in a collection of contigs representing bacterial genomes; these contigs can be binned into groups representing individual genomes based genome abundance (as deduced from read depth) and tetranucleotide frequency, a strategy that has been successfully applied in the MetaBAT tool [72].

Examples of Microbiome Approaches in Study of Foodborne Pathogens and Spoilage Organisms

A major motivation of using microbiome approaches to detect and identify foodborne pathogens is the increasing popularity of culture-independent diagnostics [73]. Surveillance networks are facing an increasing risk of losing the opportunity to obtain cultures as clinical laboratories adopt culture-independent methods. Without isolates to perform subtyping and other assays, such as antimicrobial susceptibility testing, the ability to conduct surveillance and outbreak investigation may be compromised. Metagenomics is becoming a potential solution to this challenge by characterizing the entirety of genetic materials directly from a specimen with the purpose of extracting as much information of specific pathogens. During a retrospective investigation of the 2011 outbreak of Shiga-toxigenic *Escherichia coli* (STEC) O104:H4 in Germany, microbiomes of fecal specimens obtained from patients were characterized by high-throughput sequencing [74]. From the majority of the samples, the genome of the outbreak strain was recovered by a greater than 1× coverage. This study demonstrated the potential of culture-independent, metagenomics diagnosis of foodborne pathogens from microbiologically complex clinical samples.

There are currently a limited number of microbiome studies focusing on foodborne pathogens, as compared to other areas of genomic research; however, it is worth noting that metagenomics analysis requires comprehensive reference databases of microbial genomes. From this perspective, ongoing efforts in WGS of major foodborne pathogens are setting the foundation for future application of metagenomics in public health.

Furthermore, some studies addressing questions about bacterial spoilage organisms can be easily translated to foodborne pathogen related research and are hence included in this section. Among some of the earlier microbiome studies involving foodborne pathogens is a study by Williams et al. [75] on the influence of season, irrigation, leaf age and *Escherichia coli* inoculation on the bacterial diversity in the lettuce phyllosphere. An interesting finding of this study was that even though *E. coli* presence was low (<0.001 % of the total reads) even in inoculated plants, microbial community differences could be found between plants with and without inoculation of *E. coli* O157:H7, suggesting the presence of *E. coli* affects the microbial community. Therefore, a better knowledge about the influence of pathogens on the microbiota in foods and food related environments may help to predict the presence of these pathogens, even when the level of contamination is low. Bokulich et al. [76] used a microbiome data and a Bayesian approach implemented in SourceTracker [77] for predicting routes of contamination with spoilage organisms. While most studies in the literature use sequencing reads generated with the Roche 454 and the Illumina platform, relatively few use other platforms. One study worth mentioning is the study of Hou et al. [78], which applies Pacific Biosciences SMRT sequencing perform an amplicon sequencing study based on full 16S gene length reads. This study shows the promises of newer long read technologies in microbiome research. Lastly we want to mention the study of Leonard et al. [79] in which the sensitivity of a metagenomic shotgun sequencing method of detecting contaminating Shiga toxin-producing *Escherichia coli* on bagged spinach is explored. The authors show that for proper detection at lower contamination levels (10 CFU/100 g) a (short) enrichment is still necessary for detection in the metagenomic sample.

In summary, the field of bioinformatics is rapidly evolving, as new sequencing technologies and new applications of NGS sequencing technologies are introduced, and older algorithms are improved in speed and efficiency. It is changing the way research is performed in microbiology, epidemiology and outbreak detection, and the field of foodborne pathogens.

References

1. Luscombe NM, Greenbaum D, Gerstein M. What is bioinformatics? A proposed definition and overview of the field. Methods Inf Med. 2001;40(4):346–58.
2. van Dijk EL, Auger H, Jaszczyszyn Y, Thermes C. Ten years of next-generation sequencing technology. Trends Genet. 2014;30(9):418–26. doi:10.1016/j.tig.2014.07.001.
3. Quail MA, Smith M, Coupland P, Otto TD, Harris SR, Connor TR, et al. A tale of three next generation sequencing platforms: comparison of Ion Torrent, Pacific Biosciences and Illumina MiSeq sequencers. BMC Genomics. 2012;13(1):1. doi:10.1186/1471-2164-13-341.
4. Meacham F, Boffelli D, Dhahbi J, Martin DIK, Singer M, Pachter L. Identification and correction of systematic error in high-throughput sequence data. BMC Bioinformatics. 2011;12:451. doi:10.1186/1471-2105-12-451.
5. Ross MG, Russ C, Costello M, Hollinger A, Lennon NJ, Hegarty R, et al. Characterizing and measuring bias in sequence data. Genome Biol. 2013;14(5):R51. doi:10.1186/gb-2013-14-5-r51.

6. Koren S, Phillippy A. One chromosome, one contig: complete microbial genomes from long-read sequencing and assembly. Curr Opin Microbiol. 2015;23:110–20. doi:10.1016/j.mib.2014.11.014.
7. Clark TA, Murray IA, Morgan RD, Kislyuk AO, Spittle KE, Boitano M, et al. Characterization of DNA methyltransferase specificities using single-molecule, real-time DNA sequencing. Nucleic Acids Res. 2012;40(4):e29. doi:10.1093/nar/gkr1146.
8. Langmead B, Salzberg SL. Fast gapped-read alignment with Bowtie 2. Nat Methods. 2012;9(4):357–9. doi:10.1038/nmeth.1923.
9. Li H, Durbin R. Fast and accurate short read alignment with Burrows-Wheeler transform. Bioinformatics. 2009;25(14):1754–60. doi:10.1093/bioinformatics/btp324.
10. Davis S, Pettengill JB, Luo Y, Payne J, Shpuntoff A, Rand H, Strain E. CFSAN SNP Pipeline: an automated method for constructing SNP matrices from next-generation sequence data. Peer J Comput Sci. 2015;1(12):e20–11. doi:10.7717/peerj-cs.20.
11. Katz LS, Petkau A, Beaulaurier J, Tyler S, Antonova ES, Turnsek MA, et al. Evolutionary dynamics of *Vibrio cholerae* O1 following a single-source introduction to Haiti. MBio. 2013;4(4), e00398–13.
12. Kwong JC, Mercoulia K, Tomita T, Easton M, Li HY, Bulach DM, et al. Prospective whole genome sequencing enhances national surveillance of *Listeria* monocytogenes. J Clin Microbiol. 2016;54(2):333–42. doi:10.1128/JCM.02344-15.
13. Gardner SN, Hall BG. When whole-genome alignments just won't work: kSNP v2 software for alignment-free SNP discovery and phylogenetics of hundreds of microbial genomes. PLoS One. 2013;8(12), e81760. doi:10.1371/journal.pone.0081760.
14. Gardner SN, Slezak T, Hall BG. kSNP3.0: SNP detection and phylogenetic analysis of genomes without genome alignment or reference genome. Bioinformatics. 2015;31:2877–8. doi:10.1093/bioinformatics/btv271.
15. Iqbal Z, Caccamo M, Turner I, Flicek P, McVean G. De novo assembly and genotyping of variants using colored de Bruijn graphs. Nat Genet. 2012;44(2):226–32. doi:10.1038/ng.1028.
16. Iqbal Z, Turner I, McVean G. High-throughput microbial population genomics using the Cortex variation assembler. Bioinformatics. 2013;29(2):275–6. doi:10.1093/bioinformatics/bts673.
17. Myers EW, Sutton GG, Delcher AL, Dew IM, Fasulo DP, Flanigan MJ, et al. A whole-genome assembly of *Drosophila*. Science. 2000;287(5461):2196–204. doi:10.1126/science.287.5461.2196.
18. Hernandez D, François P, Farinelli L, Osterås M, Schrenzel J. De novo bacterial genome sequencing: millions of very short reads assembled on a desktop computer. Genome Res. 2008;18(5):802–9. doi:10.1101/gr.072033.107.
19. Chevreux B, Wetter T, Suhai S. Genome sequence assembly using trace signals and additional sequence information. Presented at the computer science and biology: proceedings of the German conference on bioinformatics (GCB), vol. 99; 1999. p. 45–56.
20. Zerbino DR, Birney E. Velvet: algorithms for de novo short read assembly using de Bruijn graphs. Genome Res. 2008;18(5):821–9. doi:10.1101/gr.074492.107.
21. Bankevich A, Nurk S, Antipov D, Gurevich AA, Dvorkin M, Kulikov AS, et al. SPAdes: a new genome assembly algorithm and its applications to single-cell sequencing. J Comput Biol. 2012;19(5):455–77. doi:10.1089/cmb.2012.0021.
22. Richardson EJ, Watson M. The automatic annotation of bacterial genomes. Brief Bioinform. 2013;14(1):1–12. doi:10.1093/bib/bbs007.
23. Seemann T. Prokka: rapid prokaryotic genome annotation. Bioinformatics. 2014;30(14):2068–9. doi:10.1093/bioinformatics/btu153.
24. Glaser P, Frangeul L, Buchrieser C, Rusniok C, Amend A, Baquero F, et al. Comparative genomics of *Listeria* species. Science. 2001;294(5543):849–52. doi:10.1126/science.1063447.
25. McClelland M, Sanderson KE, Spieth J, Clifton SW, Latreille P, Courtney L, et al. Complete genome sequence of *Salmonella enterica* serovar Typhimurium LT2. Nature. 2001;413(6858):852–6. doi:10.1038/35101614.
26. Parkhill J, Dougan G, James KD, Thomson NR, Pickard D, Wain J, et al. Complete genome sequence of a multiple drug resistant *Salmonella enterica* serovar Typhi CT18. Nature. 2001;413(6858):848–52. doi:10.1038/35101607.

27. Aziz RK, Bartels D, Best AA, DeJongh M, Disz T, Edwards RA, et al. The RAST Server: rapid annotations using subsystems technology. BMC Genomics. 2008;9:75. doi:10.1186/1471-2164-9-75.
28. Delcher AL, Harmon D, Kasif S, White O, Salzberg SL. Improved microbial gene identification with GLIMMER. Nucleic Acids Res. 1999;27(23):4636–41.
29. Hyatt D, Chen G-L, Locascio PF, Land ML, Larimer FW, Hauser LJ. Prodigal: prokaryotic gene recognition and translation initiation site identification. BMC Bioinformatics. 2010;11(1):119. doi:10.1186/1471-2105-11-119.
30. Vernikos G, Medini D, Riley DR, Tettelin H. Ten years of pan-genome analyses. Curr Opin Microbiol. 2015;23:148–54. doi:10.1016/j.mib.2014.11.016.
31. Li L, Stoeckert CJ, Roos DS. OrthoMCL: identification of ortholog groups for eukaryotic genomes. Genome Res. 2003;13(9):2178–89. doi:10.1101/gr.1224503.
32. Benedict MN, Henriksen JR, Metcalf WW, Whitaker RJ, Price ND. ITEP: an integrated toolkit for exploration of microbial pan-genomes. BMC Genomics. 2014;15(1):8. doi:10.1186/1471-2164-15-8.
33. Enright AJ, Van Dongen S, Ouzounis CA. An efficient algorithm for large-scale detection of protein families. Nucleic Acids Res. 2002;30(7):1575–84.
34. Altschul SF, Gish W, Miller W, Myers EW, Lipman DJ. Basic local alignment search tool. J Mol Biol. 1990;215(3):403–10. doi:10.1006/jmbi.1990.9999.
35. Page AJ, Cummins CA, Hunt M, Wong VK, Reuter S, Holden MTG, et al. Roary: rapid large-scale prokaryote pan genome analysis. Bioinformatics. 2015;31(22):3691–3. doi:10.1093/bioinformatics/btv421.
36. Fu L, Niu B, Zhu Z, Wu S, Li W. CD-HIT: accelerated for clustering the next-generation sequencing data. Bioinformatics. 2012;28(23):3150–2. doi:10.1093/bioinformatics/bts565.
37. Felsenstein, J. Inferring phylogenies. Sunderland: Sinauer Associates, 2004.
38. Huson DH, Bryant D. Application of phylogenetic networks in evolutionary studies. Mol Biol Evol. 2006;23(2):254–67. doi:10.1093/molbev/msj030.
39. Kolaczkowski B, Thornton JW. Long-branch attraction bias and inconsistency in Bayesian phylogenetics. PLoS One. 2009;4(12), e7891. doi:10.1371/journal.pone.0007891.
40. Stamatakis A. RAxML version 8: a tool for phylogenetic analysis and post-analysis of large phylogenies. Bioinformatics. 2014;30(9):1312–3. doi:10.1093/bioinformatics/btu033.
41. Bazinet AL, Zwickl DJ, Cummings MP. A gateway for phylogenetic analysis powered by grid computing featuring GARLI 2.0. Syst Biol. 2014;63(5):812–8. doi:10.1093/sysbio/syu031.
42. Guindon S, Gascuel O. A simple, fast, and accurate algorithm to estimate large phylogenies by maximum likelihood. Syst Biol. 2003;52(5):696–704.
43. Leaché AD, Banbury BL, Felsenstein J, Nieto-Montes de Oca A, Stamatakis A. Short tree, long tree, right tree, wrong tree: new acquisition bias corrections for inferring SNP phylogenies. Syst Biol. 2015;64:1032–47. doi:10.1093/sysbio/syv053.
44. Lewis PO. A likelihood approach to estimating phylogeny from discrete morphological character data. Syst Biol. 2001;50(6):913–25.
45. Drummond AJ, Rambaut A. BEAST: Bayesian evolutionary analysis by sampling trees. BMC Evol Biol. 2007;7:214. doi:10.1186/1471-2148-7-214.
46. Schierup MH, Hein J. Consequences of recombination on traditional phylogenetic analysis. Genetics. 2000;156(2):879–91.
47. Anisimova M, Nielsen R, Yang Z. Effect of recombination on the accuracy of the likelihood method for detecting positive selection at amino acid sites. Genetics. 2003;164(3):1229–36.
48. Shriner D, Nickle DC, Jensen MA, Mullins JI. Potential impact of recombination on sitewise approaches for detecting positive natural selection. Genet Res. 2003;81(2):115–21. doi:10.1017/S0016672303006128.
49. Posada D, Crandall KA. The effect of recombination on the accuracy of phylogeny estimation. J Mol Evol. 2002;54(3):396–402. doi:10.1007/s00239-001-0034-9.
50. Didelot X, Wilson DJ. ClonalFrameML: efficient inference of recombination in whole bacterial genomes. PLoS Comput Biol. 2015;11(2), e1004041. doi:10.1371/journal.pcbi.1004041.

51. Croucher NJ, Page AJ, Connor TR, Delaney AJ, Keane JA, Bentley SD, et al. Rapid phylogenetic analysis of large samples of recombinant bacterial whole genome sequences using Gubbins. Nucleic Acids Res. 2014;43(3), e15. doi:10.1093/nar/gku1196.
52. Marttinen P, Hanage WP, Croucher NJ, Connor TR, Harris SR, Bentley SD, Corander J. Detection of recombination events in bacterial genomes from large population samples. Nucleic Acids Res. 2012;40(1), e6. doi:10.1093/nar/gkr928.
53. Hedge J, Wilson DJ. Bacterial phylogenetic reconstruction from whole genomes is robust to recombination but demographic inference is not. mBio. 2014;5(6), e02158–14. doi:10.1128/mBio.02158-14.
54. Perna NT, Plunkett G, Burland V, Mau B, Glasner JD, Rose DJ, et al. Genome sequence of enterohaemorrhagic *Escherichia coli* O157:H7. Nature. 2001;409(6819):529–33. doi:10.1038/35054089.
55. Loman NJ, Misra RV, Dallman TJ, Constantinidou C, Gharbia SE, Wain J, Pallen MJ. Performance comparison of benchtop high-throughput sequencing platforms. Nat Biotechnol. 2012;30(5):434–9. doi:10.1038/nbt.2198.
56. Quick J, Ashton P, Calus S, Chatt C, Gossain S, Hawker J, et al. Rapid draft sequencing and real-time nanopore sequencing in a hospital outbreak of Salmonella. Genome Biol. 2015;16(1):114. doi:10.1186/s13059-015-0677-2.
57. Earl D, Bradnam K, St John J, et al. Assemblathon 1: a competitive assessment of de novo short read assembly methods. Genome Res. 2011;21(12):2224–41. doi:10.1101/gr.126599.111.
58. Keim P, van Ert MN, Pearson T, Vogler AJ, Huynh LY, Wagner DM. Anthrax molecular epidemiology and forensics: using the appropriate marker for different evolutionary scales. Infect Genet Evol. 2004;4(3):205–13. doi:10.1016/j.meegid.2004.02.005.
59. Cody AJ, McCarthy ND, Jansen van Rensburg M, Isinkaye T, Bentley SD, Parkhill J, et al. Real-time genomic epidemiological evaluation of human Campylobacter isolates by use of whole-genome multilocus sequence typing. J Clin Microbiol. 2013;51(8):2526–34. doi:10.1128/JCM.00066-13.
60. Ruppitsch W, Pietzka A, Prior K, Bletz S, Fernandez HL, Allerberger F, et al. Defining and evaluating a core genome multilocus sequence typing scheme for whole-genome sequence-based typing of *Listeria* monocytogenes. J Clin Microbiol. 2015;53(9):2869–76. doi:10.1128/JCM.01193-15.
61. Venter JC, Remington K, Heidelberg JF, Halpern AL, Rusch D, Eisen JA, et al. Environmental genome shotgun sequencing of the Sargasso Sea. Science. 2004;304(5667):66–74. doi:10.1126/science.1093857.
62. Wagner Mackenzie B, Waite DW, Taylor MW. Evaluating variation in human gut microbiota profiles due to DNA extraction method and inter-subject differences. Front Microbiol. 2015;6:130. doi:10.3389/fmicb.2015.00130.
63. Sallen B, Rajoharison A, Desvarenne S, Quinn F, Mabilat C. Comparative analysis of 16S and 23S rRNA sequences of *Listeria* species. Int J Syst Bacteriol. 1996;46(3):669–74.
64. Caporaso JG, Kuczynski J, Stombaugh J, Bittinger K, Bushman FD, Costello EK, et al. QIIME allows analysis of high-throughput community sequencing data. Nat Methods. 2010;7(5):335–6. doi:10.1038/nmeth.f.303.
65. Schloss PD, Westcott SL, Ryabin T, Hall JR, Hartmann M, Hollister EB, et al. Introducing mothur: open-source, platform-independent, community-supported software for describing and comparing microbial communities. Appl Environ Microbiol. 2009;75(23):7537–41. doi:10.1128/AEM.01541-09.
66. Edgar RC. Search and clustering orders of magnitude faster than BLAST. Bioinformatics. 2010;26(19):2460–1. doi:10.1093/bioinformatics/btq461.
67. McDonald D, Price MN, Goodrich J, Nawrocki EP, Desantis TZ, Probst A, et al. An improved Greengenes taxonomy with explicit ranks for ecological and evolutionary analyses of bacteria and archaea. ISME J. 2012;6(3):610–8. doi:10.1038/ismej.2011.139.
68. Cole JR, Wang Q, Cardenas E, Fish J, Chai B, Farris RJ, et al. The Ribosomal Database Project: improved alignments and new tools for rRNA analysis. Nucleic Acids Res. 2009;37(Database issue):D141–5. doi:10.1093/nar/gkn879.

69. Mitra S, Stärk M, Huson DH. Analysis of 16S rRNA environmental sequences using MEGAN. BMC Genomics. 2011;12 Suppl 3:S17. doi:10.1186/1471-2164-12-S3-S17.
70. Li D, Liu C-M, Luo R, Sadakane K, Lam T-W. MEGAHIT: an ultra-fast single-node solution for large and complex metagenomics assembly via succinct de Bruijn graph. Bioinformatics. 2015;31(10):1674–6. doi:10.1093/bioinformatics/btv033.
71. Nurk S, Meleshko D, Korobeynikov A, Pevzner P. metaSPAdes: a new versatile de novo metagenomics assembler. 2016; arXiv:1604.03071v1.
72. Kang DD, Froula J, Egan R, Wang Z. MetaBAT, an efficient tool for accurately reconstructing single genomes from complex microbial communities. PeerJ. 2015;3, e1165. doi:10.7717/peerj.1165.
73. Cronquist AB, Mody RK, Atkinson R, Besser J, Tobin-D'Angelo M, Hurd S, et al. Impacts of culture-independent diagnostic practices on public health surveillance for bacterial enteric pathogens. Clin Infect Dis. 2012;54 Suppl 5:S432–9. doi:10.1093/cid/cis267.
74. Loman NJ, Constantinidou C, Christner M, Rohde H, Chan JZ-M, Quick J, et al. A culture-independent sequence-based metagenomics approach to the investigation of an outbreak of Shiga-toxigenic *Escherichia coli* O104:H4. JAMA. 2013;309(14):1502–10. doi:10.1001/jama.2013.3231.
75. Williams TR, Moyne A-L, Harris LJ, Marco ML. Season, irrigation, leaf age, and *Escherichia coli* inoculation influence the bacterial diversity in the lettuce phyllosphere. PLoS One. 2013;8(7), e68642. doi:10.1371/journal.pone.0068642.
76. Bokulich NA, Bergsveinson J, Ziola B, Mills DA. Mapping microbial ecosystems and spoilage-gene flow in breweries highlights patterns of contamination and resistance. Elife. 2015;4, e04634. doi:10.7554/eLife.04634.
77. Knights D, Kuczynski J, Charlson ES, Zaneveld J, Mozer MC, Collman RG, et al. Bayesian community-wide culture-independent microbial source tracking. Nat Methods. 2011;8(9):761–3. doi:10.1038/nmeth.1650.
78. Hou Q, Xu H, Zheng Y, Xi X, Kwok L-Y, Sun Z, et al. Evaluation of bacterial contamination in raw milk, ultra-high temperature milk and infant formula using single molecule, real-time sequencing technology. J Dairy Sci. 2015;98(12):8464–72. doi:10.3168/jds.2015-9886.
79. Leonard SR, Mammel MK, Lacher DW, Elkins CA. Application of metagenomic sequencing to food safety: detection of Shiga Toxin-producing *Escherichia coli* on fresh bagged spinach. Appl Environ Microbiol. 2015;81(23):8183–91. doi:10.1128/AEM.02601-15.

Chapter 5
The CGE Tool Box

Mette Voldby Larsen, Katrine G. Joensen, Ea Zankari, Johanne Ahrenfeldt, Oksana Lukjancenko, Rolf Sommer Kaas, Louise Roer, Pimlapas Leekitcharoenphon, Dhany Saputra, Salvatore Cosentino, Martin Christen Frølund Thomsen, Jose Luis Bellod Cisneros, Vanessa Jurtz, Simon Rasmussen, Thomas Nordahl Petersen, Henrik Hasman, Thomas Sicheritz-Ponten, Frank M. Aarestrup, and Ole Lund

Introduction

Human and animal health worldwide is increasingly threatened by new and re-emerging epidemics and foodborne pathogens, placing a burden on health and veterinary systems, reducing consumer confidence in food, and negatively affecting trade, food chain sustainability and food security. Rapid identification of emerging and foodborne pathogens and subsequent provision of timely insights into the modes of transmission, prevention, and control, pathogenesis, and clinical impact of such diseases is essential to reduce the impact, time, and costs of disease outbreaks.

M.V. Larsen (✉) • J. Ahrenfeldt • D. Saputra • M.C.F. Thomsen • J.L.B. Cisneros • V. Jurtz • S. Rasmussen • T.N. Petersen • T. Sicheritz-Ponten • O. Lund
Department of Systems Biology, Center for Biological Sequence Analysis, Technical University of Denmark, 2800 Kgs. Lyngby, Denmark
e-mail: metteb@cbs.dtu.dk; johah@cbs.dtu.dk; dhany@cbs.dtu.dk; mcft@cbs.dtu.dk; cisneros@cbs.dtu.dk; vanessa@cbs.dtu.dk; simon@cbs.dtu.dk; tnp@cbs.dtu.dk; thomas@cbs.dtu.dk; lund@cbs.dtu.dk

K.G. Joensen • H. Hasman • F.M. Aarestrup
Department of Microbiology and Infection Control, Statens Serum Institut, Copenhagen, Denmark
e-mail: KNJ@ssi.dk; henh@ssi.dk; fmaa@food.dtu.dk

E. Zankari • O. Lukjancenko
Department of Systems Biology, Center for Biological Sequence Analysis, Technical University of Denmark, 2800 Kgs. Lyngby, Denmark

National Food Institute, Technical University of Denmark, 2800, Kgs. Lyngby, Denmark
e-mail: east@food.dtu.dk; oklu@food.dtu.dk

X. Deng et al. (eds.), *Applied Genomics of Foodborne Pathogens*, Food Microbiology and Food Safety, DOI 10.1007/978-3-319-43751-4_5

A potential breakthrough is offered by the revolution in genome technology, leading to increasing speed and reducing costs of sequencing. As the common denominator to all pathogens and hosts, regardless of species and domain, is the presence of a genome, the ability to rapidly determine the sequence provides a common language by which data on pathogens can be compared. Such a single technology applicable to different disciplines (bacteriology, virology, parasitology) and domains (human, food, animal, environment) would facilitate global cross-cutting collaboration and information exchange (integrated surveillance), leading to rapid and coordinated responses to novel and known health threats as they emerge [1].

Conditional to this success is the capacity to generate and analyze the complex genome data in a manner that addresses clinical and public health questions reliably and timely. Thus, one of the biggest obstacles for the implementation of Whole Genome Sequence (WGS) data in clinical, animal and food microbiological laboratories is the absence of bioinformatics expertise to handle the vast amount of data. If we can provide reliable real-time bioinformatics services for frontline diagnostics, we might also be able to capture this information globally and thus create real-time global surveillance.

Center for Genomic Epidemiology (GGE) at the Technical University of Denmark was initiated in 2010 to provide a proof of concept for this. The center was funded by a grant from the Danish Council for Strategic Research. It had Prof. Frank M. Aarestrup as the coordinator and Prof. Ole Lund as the deputy coordinator.

Basically, the aim of CGE was to develop methods that use WGS data for discovering the content of a sample (typing), predict its pathogenic potential, and which antimicrobials it might be resistant towards (phenotyping). For epidemiological purposes, it was furthermore the aim to develop methods for examining the evolutionary relationship of the isolate vs. other samples. Table 5.1 provides an overview of the 16 methods that have so far been developed at CGE, and which are all made public available via easy-to-use web-services. Each method is described in more detail in the remainder of the chapter. Further, throughout the chapter the use of the web-services is exemplified with a case story employing WGS data from verotoxigenic *Escherichia coli* (VTEC) (see the boxes).

R.S. Kaas • L. Roer • P. Leekitcharoenphon
National Food Institute, Technical University of Denmark, 2800, Kgs. Lyngby, Denmark
e-mail: rkmo@food.dtu.dk; lroe@food.dtu.dk; pile@food.dtu.dk

S. Cosentino
Department of Infection Metagenomics, Genome Information Research Center, Research Institute for Microbial Diseases, Osaka University, 3-1 Yamadaoka, Suita, Osaka 565-0871, Japan
e-mail: salvocos@gen-info.osaka-u.ac.jp

Table 5.1 Overview of 16 public available web-services from CGE by Oct. 2014

Name of method	Description	URL	Publication
CSIPhylogeny	SNP-based creation of phylogenetic trees	https://cge.cbs.dtu.dk/services/CSIPhylogeny	Published Aug 2014 PMID: 25110940 [2]
KmerFinder	Species identification by co-occurring 16-mers	https://cge.cbs.dtu.dk/services/KmerFinder	Published Jan 2014 PMID: 24172157 [3]
MLST	Multilocus sequence typing	https://cge.cbs.dtu.dk/services/MLST	Published Apr 2012 PMID: 22238442 [4]
MyDbFinder	Identification of genes in user-made database	https://cge.cbs.dtu.dk/services/MyDbFinder	Published here
NDtree	Creation of phylogenetic trees	https://cge.cbs.dtu.dk/services/NDtree	Published Feb 2014 PMID: 24505344 [5]
PanFunPro	Groups homologous proteins based on functional domain content	https://cge.cbs.dtu.dk/services/PanFunPro	Published Dec 2013 [6]
PathogenFinder	Prediction of pathogenic potential	https://cge.cbs.dtu.dk/services/PathogenFinder	Published Oct 2013 PMID: 24204795 [7]
PlasmidFinder	Plasmid identification in *Enterobacteriaceae*	https://cge.cbs.dtu.dk/services/PlasmidFinder	Published Apr 2014 PMID: 24777092 [8]
pMLST	pMLST of plasmids in *Enterobacteriaceae*	https://cge.cbs.dtu.dk/services/pMLST	Published Apr 2014 PMID: 24777092 [8]
Reads2Type	Species identification on client computer	https://cge.cbs.dtu.dk/services/Reads2Type	Published Feb 2014 PMID: 24574292
ResFinder	Identification of acquired antimicrobial resistance genes	https://cge.cbs.dtu.dk/services/ResFinder	Published Nov 2012 PMID: 22782487 [9]
SerotypeFinder	WGS-based serotyping of *Escherichia coli*	https://cge.cbs.dtu.dk/services/serotypefinder	Published May 2015 PMID: 25972421 [10]
SnpTree	SNP-based creation of phylogenetic trees	https://cge.cbs.dtu.dk/services/snpTree	Published Dec 2012 PMID: 23281601 [11]
SpeciesFinder	16S rRNA-based species identification	https://cge.cbs.dtu.dk/services/SpeciesFinder	Published Feb 2014 PMID: 24574292 [12]
TaxonomyFinder	Taxonomy identification using functional protein domains	https://cge.cbs.dtu.dk/services/TaxonomyFinder	Published Feb 2014 PMID: 24574292 [12]
VirulenceFinder	Identification of virulence genes in *E. coli*	https://cge.cbs.dtu.dk/services/VirulenceFinder	Published Feb 2014 PMID: 24574290 [5]

VTEC Case Study

Verocytotoxin-producing *E. coli* (VTEC), also commonly referred to as Shiga toxin-producing *E. coli* (STEC) is a gastrointestinal pathogen, which causes disease due to production of verocytotoxins as well as several other virulence factors [13, 14]. Some VTEC cause severe infection with bloody diarrhea and at times life-threatening Hemolytic Uremic Syndrome (HUS) [15]. Around 5–10 % of VTEC infections lead to the development of HUS, and although most patients recover within a few weeks, it can be fatal or lead to permanent kidney damage. VTEC is usually contracted by ingestion of contaminated food or water, or through person-to-person contact, and it is estimated that 265,000 VTEC infections occur each year in the US [16]. In Denmark, routine typing of VTEC infections for surveillance is carried out at Statens Serum Institut (SSI). It includes serotyping (O:(K):H), which identifies the Lipopolysaccharide (O-antigen), capsular (K) antigen, and the flagellar (H) antigen. Isolates are further examined for β-glucuronidase activity, haemolysin production, and for the production of verocytotoxin, as well as for specific virulence factors, most importantly, verotoxin 1 (*vtx1*), verotoxin 2 (*vtx2*) and intimin (*eae*), which are detected by DNA hybridization, and further subtyping of the verocytotoxins is carried out by PCR. Isolates of the same serotype with similar toxin profiles and phenotypic features that are considered to be potential outbreak isolates are further subjected to PFGE typing for comparison. Due to the many different analysis that are necessary for accurate routine typing and surveillance, it is time-consuming, laborious and costly. Thus, as a proof-of-concept that WGS-based typing of VTEC could be an attractive alternative, real-time WGS-based typing of VTEC were performed during 7 weeks, in parallel to the routine typing carried out at SSI. The study included a set of 46 suspected VTEC isolates and has previously been described [5]. Throughout this chapter, the same set of suspected VTEC isolates are used to exemplify the use and output of selected CGE web-services.

Prokaryotic Taxonomy

From a pragmatic point of view "the ultimate goal of taxonomy is to construct a classification that is of operative and predictive use for any discipline in microbiology and that is also essentially stable" [17]. Most taxonomists agree that phylogeny—the study of the evolutionary history of organisms—should be the underlying basis of taxonomy. Historically, the first attempts on bacterial classification were based on morphology, later the phylogenetic reconstructions were based on physiological properties. A milestone in classification of prokaryotes was the introduction of 16S rRNA sequence data [18] and it has dominated molecular taxonomy since. Tremendous amounts of 16S rRNA gene sequence data are available in public databases [19, 20]. Several concerns about its use have, however, been raised. These include low resolution [21, 22], the presence of several, and sometimes-different 16S rRNA genes in some genomes [23], and the fact that the 16S rRNA gene only represents a tiny fraction of microbial genomes [24]. The introduction of WGS data enables alternative approaches for prokaryotic classification that utilize a larger portion of the genome. At CGE, a number of methods using WGS data have been implemented. Some examine only the 16S rRNA gene (SpeciesFinder and Reads2Type), while others take a larger portion of the genomes into account (TaxonomyFinder and KmerFinder) [25].

SpeciesFinder

SpeciesFinder predicts prokaryotic species based on the 16S rRNA gene. For this purpose, it uses a database of 16S rRNA genes isolated from reference genomes with annotated species [25]. The predicted species of a query genome is selected as the annotated species of the reference genome with the most similar 16S rRNA gene. The SpeciesFinder web-service exemplifies the most basic version of the generally simple CGE user interface (Fig. 5.1). The user only has to select the input file (short sequence reads in FASTQ format or a draft genome in FASTA format) containing the DNA sequence of the query isolate and click the "Submit" button.

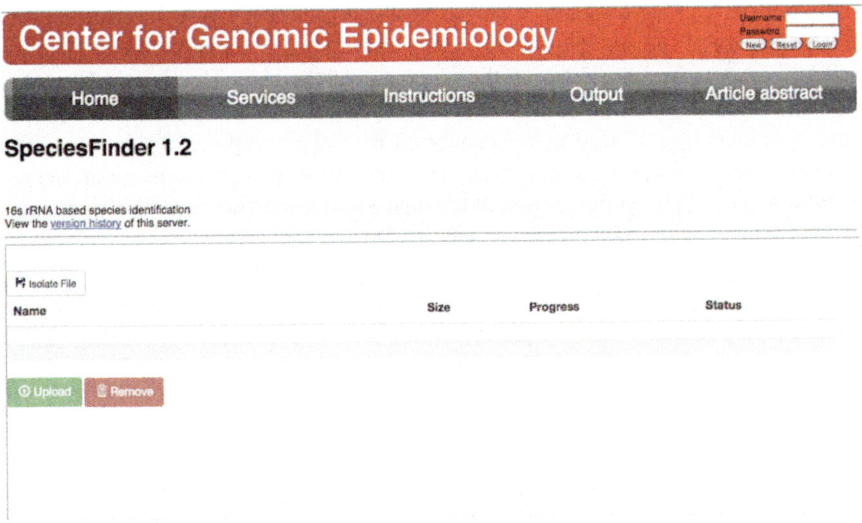

Fig. 5.1 User interface of the SpeciesFinder web-service. Using the "Browse" button the input file (short sequence reads in FASTQ format or a draft genome in FASTA format) containing the DNA sequence of the query isolate is selected. Then, the "Submit" button is clicked

VTEC Case Study: Identifying the Species Using SpeciesFinder

One of the suspected VTEC isolates (C757-12) [5] was run through the SpeciesFinder web-service to confirm the species as *E. coli*. Like the input page of the web-service, the output page is very simple (Fig. 5.2). Besides the predicted species and the GenBank accession number of the reference genome on which the prediction is based, the confidence level of the result is marked as PASS or FAIL; if the prediction is based on a similarity between the 16S rRNA gene of the SpeciesFinder database and the 16S rRNA gene of the query genome above 98 % identity on nucleotide level, the confidence of result is listed as "PASS". Otherwise it is listed as "FAIL".

SpeciesFinder is available at https://cge.cbs.dtu.dk/services/SpeciesFinder.

Reads2Type

Similar to SpeciesFinder, the 16S rRNA gene forms the basis of the Reads2Type method. However, instead of examining similarity across the entire gene, the method employs a small, pre-made database of species-specific 50-mers (stretches of DNA with the length of 50 nucleotides) from within the gene. Further, for the *Enterobacteriaceae* family, the *GyrB* gene is used as the species-specific marker gene, not the 16S rRNA gene. When using the Reads2Type web application, this small database of species-specific 50-mers is automatically transferred into memory and all computations are done on the computer of the user. This is an advantage, since it means that the much larger data amounts that the query genome sequence data represents do not need to be transferred from the computer of the user to the central server. Besides minimizing data transfer, bottleneck problems on the server are also avoided. The minimization of the data transfer may be particularly advantageous for users with limited Internet access [25].

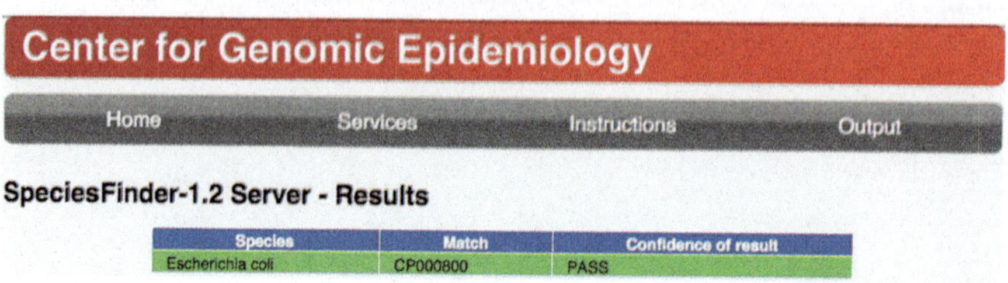

Fig. 5.2 The SpeciesFinder output when the suspected VTEC isolate C757-12 [5] is used as input. The output includes the predicted species (Species), the accession number of the reference genome on which the prediction is based (Match), and the level of confidence of the prediction (Confidence of result)

Reads2Type is available at https://cge.cbs.dtu.dk/services/Reads2Type.

TaxonomyFinder

The pan-genome of a given taxonomic group of genomes (phylum, genus, species) consists of a set of conserved genes, genes that are present in some, but not all genomes, and genes that are specific for particular strains. The typical approach for taxonomy prediction is using evolutionary conserved genes; such as 16S rRNA or a set of 'housekeeping' genes as in MLST (see below). However, taxonomic classification can also be performed using taxa-group specific proteins, which is the approach applied by TaxonomyFinder.

The TaxonomyFinder specific protein database was created using PanFunPro (Pan-genome analysis based on functional profiles) [6], homology detection, and a protein annotation tool. PanFunPro can be used for core-, pan- and accessory genome analysis, such as estimation of life's set of core genes, prediction of chromosome-specific families [26], analysis of differences between probiotic and pathogenic strains [6], as well as estimation of taxonomy-group specific proteins. Briefly, a set of proteins from a number of prokaryotic genomes are searched for functional domains using the InterProScan software [27] against the three Hidden Markov Model (HMM) collections: PfamA, SuperFamily and TIGRFAM. Subsequently, non-repeating and non-overlapping functional domains within each protein are combined into functional profiles, using the information of one database at a time, with respect to the order of scans. Homologous proteins are grouped into protein families, based on functional profiles. Next, taxa-specific profiles are predicted. A profile is considered to be specific, if it is 100 % conserved among the set of query genomes (genomes within a taxonomic group), and absent in the rest of the analyzed genomes. However, this approach may not be feasible if the number of members in the taxonomic group is very high, such as in the *Firmicutes* and *Proteobacteria* phyla, or *Escherichia* genus. Under these circumstances, the requirement is lowered, meaning that profiles remain specific to the taxonomic group, but may be missing in several genomes within the group.

TaxonomyFinder implements taxonomy prediction on species and phylum level. The database includes 33 phylum-specific and 1242 species-specific profile sets. Additionally, TaxonomyFinder provides protein annotation for the submitted isolate based on functional profiles.

PanFunPro is available at https://cge.cbs.dtu.dk/services/PanFunPro.

TaxonomyFinder is available at https://cge.cbs.dtu.dk/services/TaxonomyFinder.

KmerFinder

In their groundbreaking paper from 1977, Woese and Fox uncovered Archea as a separate branch in the tree of life [28]. As a measure of genetic relatedness, they used the number of co-occurring kmers in 16S (18S) rRNA genes. Kmers are stretches of DNA with the length of k nucleotides (Woese and Fox used the term oligonucleotides). Taking advantage of the availability of complete prokaryotic genomes, KmerFinder uses a similar approach for identifying the species, but extends the analysis to kmers across the entire genome. More specifically, KmerFinder relies on a database of reference genomes with annotated species [25] that are each split into overlapping 16-mers with step-size one. This means that if the first 16-mer of a reference genome is initiated at position N and ends at position N + 15, then the next 16-mer is initiated at position N + 1 and ends at position N + 16 etc. To reduce the size of the final 16-mer database, only 16-mers with the prefix ATGAC are kept. For the prediction of the species of a query genome, the genome is likewise split into overlapping 16-mers and the species is predicted to be identical to the species of the reference genome with which it has the highest number of 16-mers in common regardless of position.

VTEC Case Study: Identifying the Species Using KmerFinder

The suspected VTEC isolate (C757-12) [5] was run through the KmerFinder web-service using the scoring method "winner takes it all". KmerFinder offers two different scoring schemes: "standard" and "the winner takes it all". In the standard scoring scheme, all identical Kmers between the query sequence and the reference genomes are reported and statistics are calculated based on this. When choosing "the winner takes it all" scoring scheme, the output for the top hit (the reference genome in which the highest number of identical 16-mers with the query sequence was found) is the same as for the standard scoring scheme. But for the following significant hits, only 16-mers that were not found before are counted. This scoring scheme leads to the indication of whether or not (and to which extent) the query sequence is chimeric—with two or more origins.

Figure 5.3 shows the KmerFinder output page. The "Hit" is the genome in the reference database with which the query genome (the suspected VTEC isolate, C757-12) has most co-occurring 16-mers. Hence, according to KmerFinder, the predicted species is *E. coli*.

For another suspected VTEC isolate, C484-12 [5], KmerFinder found that the isolate was actually a *Morganella Morganii*. This was later confirmed and the isolate was excluded from the remainder of the study.

KmerFinder is available at https://cge.cbs.dtu.dk/services/KmerFinder.

Performance of Methods for Species Identification

The performances of the above-mentioned four methods for species identification have been evaluated in terms of accuracy and speed [25]. More than 11,000 isolates covering 159 genera and 243 species were used in the evaluation. Datasets of both

KmerFinder 2.0 results:

Hit	Score	z-score	Query Coverage [%]	Template Coverage [%]	Depth	Total Query Coverage [%]	Total Template Coverage [%]	Total Depth
Escherichia coli, Escherichia coli O103:H2, Escherichia coli O103:H2 str. 12009 get sequence	10207	410.3	18.08	93.19	18.62	18.08	93.19	18.62

Fig. 5.3 The KmerFinder output when the suspected VTEC isolate C757-12, with serotype O103:H2 [5], is used as input. The columns of the output table contain, among others, the name of the reference genome with which the query genome has the highest number co-occurring 16-mers and a link to the sequence of the genome (first column; Hit) The remaining columns contain statistics on the 16-mers used for the comparison. A full description of the content of all columns can be found here: http://cge.cbs.dtu.dk/services/KmerFinder/output.php

short sequence reads and assembled draft genomes were included. The results indicated that methods, which only sample chromosomal, core genes (e.g. SpeciesFinder and Reads2Type) had difficulties in distinguishing closely related species. As an example, SpeciesFinder had problems distinguishing *Yersinia pestis* from *Yersinia pseudotuberculosis* and *Mycobacterium tuberculosis* from *Mycobacterium bovis*. Overall, the performances of SpeciesFinder and Reads2Type were found to be similar, ranging from 76 to 87 % correct species identification, when tested on the three different evaluation sets. TaxonomyFinder, on the other hand, had a higher performance, ranging from 85 to 95 % correct species identification. KmerFinder had the highest performance with 93–97 % correct identifications. The species that all methods had problems distinguishing were typically within the *Bacillus* genus or *Escherichia coli—Shigella* spp. mix-ups. Rather than pointing to flaws in the methods, these misclassifications are likely to highlight sub-optimal conventional classification: Species belonging to the *Bacillus cereus* group are notoriously difficult to distinguish, as they are genetically very similar. It has hence been suggested that all members of the *B. cereus* group (including *B. cereus*, *Bacillus anthracis*, and *Bacillus thuringiensis*) should be considered to be *B. cereus* and only subsequently differentiated by their content of plasmids [29]. Likewise, although *Shigella* spp. has for many years been considered a sub-strain of *E. coli* and the separation is mainly historical [30, 31], the practical implications of renaming it are considered insurmountable.

The speed of the four methods for species identification was likewise tested on both assembled draft genomes and short sequence reads (see Table 5.2). Since the actual speed experienced by the user will depend on, for instance, the network bandwidth capacity of their computer and the number of jobs queued at the server, it is the relative speed of the different methods in comparison to each other that should be noted, not the absolute speed. KmerFinder was found to be the fastest method, while TaxonomyFinder was the slowest. In contrary to the other methods, TaxonomyFinder does not work on the nucleotide sequence directly, but rather on the proteome, utilizing functional protein domain profiles for the species prediction. Hence, in return for the extra time, the user is rewarded with an annotated genome.

Table 5.2 Speed of four methods for whole genome-based species identification

Method	Speed on draft genomes (mm:ss)	Speed on short reads (mm:ss)
SpeciesFinder	00:13	3:14
Reads2Type	NA[a]	1:20
TaxonomyFinder	11:33	NA[a]
KmerFinder	00:09	03:10

[a]Reads2Type only takes short sequence reads as input, while TaxonomyFinder only takes assembled draft genomes as input

Subtyping

Once the species of the organism of interest has been identified, the next step is typically to identify the strain. A number of methods have been proposed—and are in use—for the purpose of differentiating microorganisms beyond the level of species or subspecies. Some of these methods are phenotype-based, e.g., phage-typing and serotyping, while others are founded on the genomes of the organisms. In 1998 a scheme for subtyping on the basis of internal nucleotide sequences of a small number of housekeeping genes was proposed for *Neisseria meningitidis* [32]. Unique sequences (alleles) for each housekeeping gene (locus) are assigned a random integer number and a unique combination of alleles at each locus, an "allelic profile", defines the sequence type (ST). The procedure is called multilocus sequence typing (MLST) and has been adopted to close to 100 additional microorganisms besides *N. meningitidis*. These additional microorganisms are mainly bacteria, but MLST schemes for fungal species have also been developed. The MLST allele sequences and ST profile tables are stored in curated databases hosted at different sites around the world, and made collectively available via the pubMLST site (http://pubmlst.org). One of the great advantages of MLST is that it is standardized and the nucleotide sequence of a particular allele of a particular locus is unambiguous, thus requiring a minimum of subjective interpretation by the person carrying out the analysis. For a handful of species, e.g., *Escherichia coli* [33, 34] and *Clostridium difficile* [35, 36], different groups have developed different MLST schemes, each employing a slightly different set of loci. But besides these inexpedient causes of confusion, MLST and the sequence types provide clinical microbiologists, food safety authorities and everyone else working with microorganisms a standardized way of performing subtyping and naming the bacteria. The latter is also extremely useful for communicative purposes.

The MLST Web-Service

Due to the above-mentioned advantages of MLST, it was for several years considered the gold standard of typing, even though it was traditionally carried out in a time-consuming and expensive manner. With the advent of next generation sequencing technologies, other typing schemes that take a larger proportion of the genome into account are expected to become predominant; e.g., SNP, wgMLST or, cgMSLT based. Nevertheless, the first web-service made available by CGE was one that could perform MLST on the basis of WGS data [4]. The purpose of developing this web-service was less to confirm the importance of MLST as a reference genome typing method, but rather to enable comparison of isolates based on WGS data with those analyzed earlier by more traditional methods. In other words, the purpose was to provide backwards compatibility, while in parallel using WGS-based MLST as a spearhead for implementation of routine-use of WGS.

For use by the CGE MLST web-service, all MLST databases are automatically downloaded from the pubMLST.org site once a week. As input, the MLST web-service can receive either short sequence reads or assembled draft genomes. In the case of short sequence reads, they are assembled to draft genomes before the analysis [4]. Applying the user-specified MLST scheme to the input data, the best-matching MLST alleles is then found using BLAST [37]. Finally, the sequence type is determined by the combination of identified alleles. Currently (Oct. 2014), 116 different schemes are available for bacterial and fungal species, and more are being added, as they are developed and incorporated in pubMLST.org.

VTEC Case Study: Identifying Sequence Type Using the MLST Web-Service

A suspected VTEC isolate (C770-12) [5] was run through the MLST web-service using the *E. coli* #1 MLST scheme. Figure 5.4 shows the output of the service. Below the listed predicted Sequence Type, a table shows the best-matching allele in the MLST database for each of the seven loci, along with information on the quality of the match.

By clicking the "extended output" button it is possible to examine the alignment between each of the MLST alleles and the corresponding sequences in the query sequence. Figure 5.5 shows the alignment of the *reca* locus, in which one gap occurs in the query sequence. All the suspected VTEC isolates were sequenced on an IonTorrent PGM sequencer, which is known to produce a large number of false-positive indels.

The remaining "Finder-tools" described below also all have the option to examine the actual alignments by selecting the extended output format.

*HSP: High-scoring Segment Pair. BLAST term that refers to the length of the alignment between the allele from the MLST database and the corresponding nucleotide sequence in the query genome.

The MLST web-service is currently (end-2015) the second-most used web-service provided by CGE (the most used web-service being ResFinder). From Sep. 2012 to Oct. 2015, the service was used more than 50,000 times in total. Examining the literature citing the CGE MLST web-service indicates that the service has most often been used for typing *E. coli*, *S. enterica*, and *S. aureus* (for examples see [3, 38–40]).

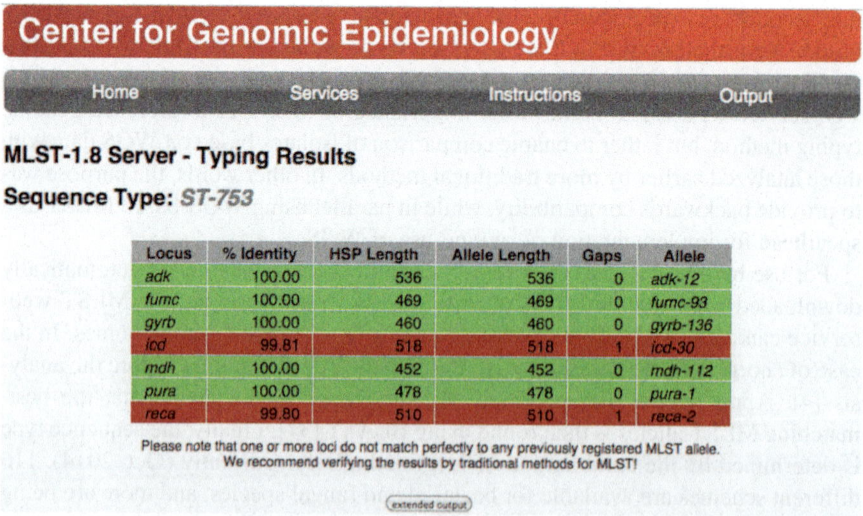

Locus	% Identity	HSP Length	Allele Length	Gaps	Allele
adk	100.00	536	536	0	adk-12
fumc	100.00	469	469	0	fumc-93
gyrb	100.00	460	460	0	gyrb-136
icd	99.81	518	518	1	icd-30
mdh	100.00	452	452	0	mdh-112
pura	100.00	478	478	0	pura-1
reca	99.80	510	510	1	reca-2

Fig. 5.4 Output from the MLST web-service when the suspected VTEC isolate, C770-12 [5], was run through the service using the *E. coli* #1 MLST scheme. Rows describing perfect matches between alleles in the database and the query sequence are colored *green*. In perfect matches the % identity is 100, the HSP* length equals the allele length, and there are no gaps. Rows describing imperfect matches are *red*

```
reca: WARNING, Identity: 99.80%, HSP/Length: 510/510, Gaps: 1,  Best Match: reca-2

MLST allele seq:    cgcacgtaaactgggcgtcgatatcgacaacctgctgtgctcccagccggacaccggcga
Hit in genome:      cgcacgtaaactgggcgtcgatatcgacaacctgctgtgctcccagccggacaccggcga

MLST allele seq:    gcaggcactggaaatctgtgacgccctggcgcgttctggcgcagtagacgttatcgtcgt
Hit in genome:      gcaggcactggaaatctgtgacgcc-tggcgcgttctggcgcagtagacgttatcgtcgt

MLST allele seq:    tgactccgtggcggcactgacgccgaaagcggaaatcgaaggcgaaatcggcgactctca
Hit in genome:      tgactccgtggcggcactgacgccgaaagcggaaatcgaaggcgaaatcggcgactctca

MLST allele seq:    catgggccttgcggcacgtatgatgagccaggcgatgcgtaagctggcgggtaacctgaa
Hit in genome:      catgggccttgcggcacgtatgatgagccaggcgatgcgtaagctggcgggtaacctgaa

MLST allele seq:    gcagtccaacacgctgctgatcttcatcaaccagatccgtatgaaaattggtgtgatgtt
Hit in genome:      gcagtccaacacgctgctgatcttcatcaaccagatccgtatgaaaattggtgtgatgtt

MLST allele seq:    cggtaacccggaaaccactaccggtggtaacgcgctgaaattctacgcctctgttcgtct
Hit in genome:      cggtaacccggaaaccactaccggtggtaacgcgctgaaattctacgcctctgttcgtct

MLST allele seq:    cgacatccgtcgtatcggcgcggtgaaagagggcgaaaacgtggtgggtagcgaaacccg
Hit in genome:      cgacatccgtcgtatcggcgcggtgaaagagggcgaaaacgtggtgggtagcgaaacccg

MLST allele seq:    cgtgaaagtggtgaagaacaaaatcgctgcgccgtttaaacaggctgaattccagatcct
Hit in genome:      cgtgaaagtggtgaagaacaaaatcgctgcgccgtttaaacaggctgaattccagatcct

MLST allele seq:    ctacggcgaaggtatcaacttctacggcga
Hit in genome:      ctacggcgaaggtatcaacttctacggcga
```

Fig. 5.5 Pairwise-alignment of the *reca-2* allele from the MLST database and the corresponding sequence in the C770-12 isolate

The MLST web-service is available at https://cge.cbs.dtu.dk/services/MLST.

Serotype

Serotyping has since its development in the 1940s become the gold standard for typing of several important pathogens, e.g. *E. coli* and *Salmonella*. Classical serotyping relies on serological detection of antigenic surface structures.

Serotyping of *E. coli*

For *E. coli* the most important antigens in typing are the somatic lipopolysaccharide O-antigen and the flagellar H-antigen. In order to transform this classical, phenotypic typing method into the WGS era, SerotypeFinder was constructed for WGS-based serotyping of *E. coli* [10]. The tool utilizes the O-antigen processing genes of *wzx*, *wxy*, *wzm*, and *wzt* to predict O-types and the flagellin genes *fliC*, *flkA*, *fllA*, *flmA*, and *flnA* for prediction of H-types. The SerotypeFinder database includes gene variants covering all 53 known H-types as well as all 188 known O-types, with the exception of O14 and O57.

The SerotypeFinder outputs O-types on basis of O-type specific gene variants, either by the combination of the variants detected by *wzx* and *wzy*, or by *wzm* and *wzt*. With a few exceptions, the two genes, e.g. *wzx/wzy*, output the same O-type, whereas in some cases the O-type will be predicted from just one of these gene variants. The H-type is predicted in SerotypeFinder by *fliC* gene variants alone, when this gene is the only flagellin gene present in the genome that is examined, whereas in cases of both a *fliC* and a non-*fliC* (*flkA*, *fllA*, *flmA*, *flnA*) gene being present, the non-*fliC* is set to predict the phenotype.

The SerotypeFinder is very robust and provides results directly comparable to the conventional serotyping of *E. coli*. In addition, it offers H-typing of all non-motile *E. coli* as well as O-typing of some rough strains that cannot be serotyped by conventional serotyping.

The SerotypeFinder is available at: https://cge.cbs.dtu.dk/services/SerotypeFinder.

Serotyping of *Salmonella*

Serotyping has likewise traditionally been a cornerstone in the surveillance of *Salmonella*. As for *E. coli*, the monitored antigens are the lipopolysaccharide O-antigen (encoded by the *rfb* gene cluster) and the flagellar H-antigen (encoded by the *fliC* and *fliB* genes). Researchers at the Center for Food Safety, University of

Georgia have developed a method, which is based on mapping of the raw reads from a sequencing run to curated databases of alleles of the *rfb* gene cluster and the *fliC* and *fliB* genes for WGS-based serotyping of *Salmonella* [41]. Although not having been involved in the development of the SeqSero web-service, CGE hosts the service, which is able to use raw sequence reads as well as assembled draft genomes as input.

The SeqSero web-service is available at: https://cge.cbs.dtu.dk/services/SeqSero and https://www.denglab.info/SeqSero.

Plasmids

Just as clones of bacteria might spread and are important to track, e.g., to find the source of an outbreak, so might plasmids spread horizontally among bacteria, conferring specific properties to their hosts. For the molecular epidemiological investigations of the major plasmid incompatibility groups among *Enterobacteriaceae*, plasmid-typing methods have been developed [42]. The initial method was developed to detect the replicons (part of the origin of replication and/or the replicase gene) of plasmids of the 18 major incompatibility (Inc) groups found in *Enterobacteriaceae*, but was extended and now contains 25 different replicons. Based on a database of 116 replicon sequences extracted from 559 plasmids, the PlasmidFinder method employs a BLAST-based search engine similar to the CGE MLST implementation for identification of plasmids [8]. For plasmid multilocus sequence typing (pMLST), a weekly updated database is furthermore generated from www.pubmlst.org/plasmid and integrated into a separate web-service named pMLST.

VTEC Case Study: Identifying Plasmids Using PlasmidFinder
Results of the PlasmidFinder web-service, when the isolate C757-12 [5] is used as input is shown in Fig. 5.6. Two plasmids (or actually replicons) were found: *I1* and *FIB(AP001918)*.

PlasmidFinder is available at https://cge.cbs.dtu.dk/services/PlasmidFinder.

The pMLST web-service is available at https://cge.cbs.dtu.dk/services/pMLST.

Phenotyping

Once the isolate is identified with adequate resolution, its potential phenotype can be investigated. For this purpose, we have developed a number of methods that typically search the input genome for the presence of particular genes that, if expressed at adequate levels, are likely to result in a particular phenotype of the isolate.

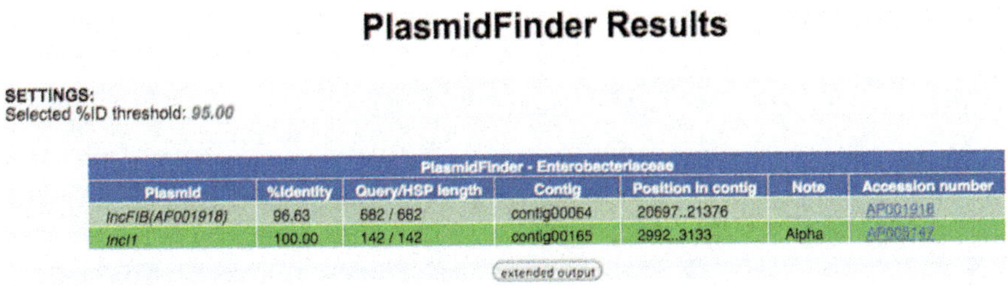

PlasmidFinder Results

SETTINGS:
Selected %ID threshold: 95.00

PlasmidFinder - Enterobacteriaceae						
Plasmid	%Identity	Query/HSP length	Contig	Position in contig	Note	Accession number
IncFIB(AP001918)	96.63	682 / 682	contig00064	20697..21376		AP001918
IncI1	100.00	142 / 142	contig00165	2992..3133	Alpha	AP005147

extended output

Fig. 5.6 Results of the PlasmidFinder web-service. Note that it is the identified replicons (part of the origin of replication and/or the replicase gene) that are reported, not the entire plasmids

ResFinder

If the isolate is a pathogen, it is of importance to find out how it can be treated. To this end, a method for identification of acquired antimicrobial resistance genes was developed. The method is called ResFinder [9]. A major effort was initially put into compiling a curated database, based on public databases as well as on scientific papers. The database contains genes for the 13 major antimicrobial agent groups (Aminoglycosides, Beta-lactamases, Fluoroquinolone, Fosfomycin, Fusidic Acid, Glycopeptides, Macrolide-Lincosamide-StreptograminB, Nitroimidazole, Phenicol, Rifampicin, Sulfonamide, Tetracycline, and Trimethoprim) and is updated continuously. Query genomes are examined for the presence of any of these genes using the BLAST-based search engine.

> **VTEC Case Study: Identifying Acquired Antibiotic Resistance Genes with ResFinder**
> All the suspected VTEC isolates were examined for the presence of acquired antibiotic resistance genes. Overall, the isolates only contained very few of these genes. Figure 5.7 shows the results for C659-12 [5], which contained three genes known to confer resistance towards aminoglycosides.

Concerns have been raised that an assigned genotype may not always correspond to the actual phenotype, for instance due to mutations outside a particular gene, but affecting the expression of the gene product. A study was therefore conducted to compare antimicrobial resistance geno- and phenotypes. A surprisingly high concordance (99.74 %) was found between phenotypic and predicted antimicrobial susceptibility. Although the results were promising, it should be noted that the study was conducted in a population with relatively low levels of resistance, and lower levels of concordance may be found in other populations. It should likewise be noted that ResFinder is only able to discover antimicrobial resistance due to acquired antimicrobial resistance genes, not, e.g., point mutations in chromosomal genes. Nevertheless, it was concluded that genotyping using whole-genome sequencings is

ResFinder-2.1 Server - Results

Aminoglycoside						
Resistance gene	%Identity	Query/HSP length	Contig	Position in contig	Predicted phenotype	Accession number
strB	99.52	837 / 837	contig00220	1080..1912	Aminoglycoside resistance Alternate name; aph(6)-Id	M96392
strA	100.00	804 / 804	contig00220	277..1080	Aminoglycoside resistance Alternate name; aph(3")-Ib	AF321551
aadA1	100.00	789 / 789	contig00096	5041..5829	Aminoglycoside resistance	JQ480156

Beta-lactam
No resistance genes found.

Fluoroquinolone
No resistance genes found.

Fosfomycin
No resistance genes found.

Fusidic Acid
No resistance genes found.

MLS - Macrolide, Lincosamide and Streptogramin B
No resistance genes found.

Nitroimidazole
No resistance genes found.

Fig. 5.7 An extract of the result from the ResFinder service when the C659-12 isolate is used as input. Besides the name of the resistance genes found (Resistance gene), the %identity between the gene in the ResFinder database and the corresponding sequence in the input isolate is shown (%identity). Also shown is the Query/HSP Length, where Query length is the length of the best matching resistance gene in the database, while HSP length is the length of the alignment between the best matching resistance gene and the corresponding sequence in the genome (also called the high-scoring segment pair (HSP)). The name of the contig and the position in the contig is also shown. Finally, the predicted phenotype based on the identified resistance gene is shown, as is the reference Genbank accession number according to NCBI of the resistance gene in the database

a realistic alternative to surveillance based on phenotypic antimicrobial susceptibility testing [43].

ResFinder is currently the most frequently used web-services provided by CGE having been used almost 60,000 times by Oct. 2015, since its publication in end-2012. A review of the literature citing ResFinder shows that the method has so far mainly been used for identifying antimicrobial resistance genes in gram-negative bacteria, e.g., *K. pneumoniae* (for instance [44]), *E. coli* (see for instance [45]), and *S. enterica* (see for instance [46]).

ResFinder is available at https://cge.cbs.dtu.dk/services/ResFinder.

MyDbFinder

To accommodate that researchers may have interests in particular sets of genes, for which there is no in-house CGE databases, a special version of ResFinder, called MyDbFinder, was developed. Using this service, the user can generate their own database containing genes of interest, for which the program should search. The database must contain the DNA sequences of the genes that the user wishes to

```
>Seq1
ACTCGCGATCCGCATAGCGCATCGCATG
>Seq2 optional comment
ATGAAAACAATGATTTATCCTCACCAATATAATTATATCAGATCGGTTATT
TATGCGGCAATGATTTATCCTCACCAATGATGAGAGAGCAGATACTCTTTG
AACAAAGAAATTGAAGCAATACTTAATAAATTT
```

Fig. 5.8 Two DNA sequences in FASTA format

search for. As ResFinder, MyDbFinder uses BLAST to identify the genes in the query WGS data and outputs the best matching genes from the user's database. It is possible to select different settings depending on how strict an output is wanted.

How to Make a Database for MyDbFinder

The database should be made in a text editor (Notepad, TextEdit or equivalent) and must consist of DNA sequences in FASTA format. A sequence in FASTA format begins with a header, which is a single line description, followed by lines of sequence data in single-letter nucleotide code (A, T, C, and G). The header line always starts with a ">" (greater than) symbol, which distinguishes this line from the lines containing the sequence data. Note that empty lines are not accepted in FASTA files.

Figure 5.8 exemplifies two DNA sequences in FASTA format:

When making a database the user should be aware that MyDbFinder only shows the first word of the header (the characters until the first space) for outputted genes, thus different genes/sequence names without spaces are recommended.

MyDbFinder is available at https://cge.cbs.dtu.dk/services/MyDbFinder.

VirulenceFinder

The same BLAST-based methodology as above is also used for prediction of virulence factors in verotoxigenic *E. coli* on the basis of a database of known *E. coli* virulence genes [5]. In the study for which this VirulenceFinder tool was developed, it was used to examine 46 suspected VTEC isolates (the same isolates that are used throughout this chapter to exemplify the output of CGE web-services). VirulenceFinder quickly and accurately detected *eae*, *ehxA*, and *vtx* genes and was in addition more robust in assigning correct *vtx* subtypes than routine typing. Although poor sequencing quality and overall low coverage for some isolates caused VirulenceFinder to miss the detection of a few genes, it overall detected the presence of many other important virulence genes, thus giving much more information on the virulence profiles of the isolates than was obtained by routine typing [5].

VTEC Case Study: Identifying Virulence Factors Using VirulenceFinder

The suspected VTEC isolates were examined for the presence of known virulence genes using VirulenceFinder. Figure 5.9 shows the results when using the isolate C892-12 [5] as input. Routine typing had assigned this isolate as *vtx2d*, while the subtype found by VirulenceFinder was *vtx2g* (reported in the column "protein function"). Retyping at SSI confirmed the subtype detected by VirulenceFinder. In a few other instances, where results obtained by routine typing and VirulenceFinder were not in agreement, poor WGS data quality was usually the cause.

VirulenceFinder is available at https://cge.cbs.dtu.dk/services/VirulenceFinder.

PathogenFinder

Whereas VirulenceFinder looks for the presence of known virulence factors previously described in the literature, Andreatta et al. took a radically different approach for determining the pathogenic potential of an organism [47]. In the study from 2010,

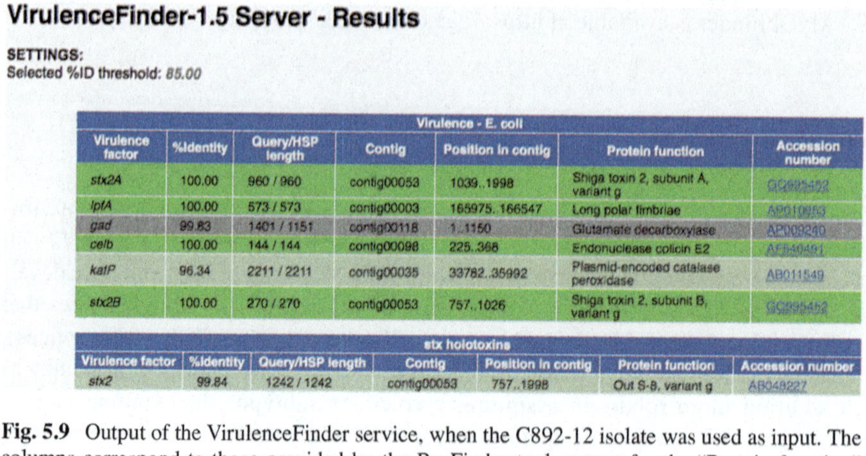

VirulenceFinder-1.5 Server - Results

SETTINGS:
Selected %ID threshold: *85.00*

Virulence - E. coli						
Virulence factor	%Identity	Query/HSP length	Contig	Position in contig	Protein function	Accession number
stx2A	100.00	960 / 960	contig00053	1039..1998	Shiga toxin 2, subunit A, variant g	GQ895452
lpfA	100.00	573 / 573	contig00003	165975..166547	Long polar fimbriae	AP010953
gad	99.83	1401 / 1151	contig00118	1..1150	Glutamate decarboxylase	AP009240
celb	100.00	144 / 144	contig00098	225..368	Endonuclease colicin E2	AF540491
katP	96.34	2211 / 2211	contig00035	33782..35992	Plasmid-encoded catalase peroxidase	AB011549
stx2B	100.00	270 / 270	contig00053	757..1026	Shiga toxin 2, subunit B, variant g	GQ895452

stx holotoxins						
Virulence factor	%Identity	Query/HSP length	Contig	Position in contig	Protein function	Accession number
stx2	99.84	1242 / 1242	contig00053	757..1998	Out S-8, variant g	AB048227

Fig. 5.9 Output of the VirulenceFinder service, when the C892-12 isolate was used as input. The columns correspond to those provided by the ResFinder tool, except for the "Protein function" column

genomes from *gamma-proteobacteria* were first grouped into those originating from pathogenic vs. non-pathogenic bacteria. Next, the genomes were examined for the presence of gene families that were statistically associated with being found in either the pathogenic or non-pathogenic groups. To the best of our knowledge, this is the first example of the use of machine learning techniques for determining the phenotype based on whole genome sequences. The method has later been extended to be applicable for all species of bacteria and made publicly available as the PathogenFinder method [7]. Since the method relies on groupings of proteins, without considering their annotated function (or even if they have any) or known involvement in pathogenicity, it can also aid the discovery of novel pathogenicity factors.

PathogenFinder is available at https://cge.cbs.dtu.dk/services/PathogenFinder.

Phylogeny

Nucleotide sequences have long been used to classify the species and taxonomy of bacteria and other organisms. But until recently it was only a few genes, including the 16s rRNA gene, that were used for making phylogenies. However, since the price of whole genome sequencing has gone down, whole genome based phylogeny has become increasingly used for both typing of bacteria and for disease outbreak detection. Previously phylogeny was mostly used to divide samples into families and species. But if the method is exact enough, it can even be used to follow and detect disease outbreaks. If, for example, WGS data from a number of samples from different patients are available, and the strains only differ by a few nucleotides, the strains are likely to have originated from the same strain, and hereby it can most likely be concluded that all the patients were contaminated from the same source. Another option that becomes possible when WGS data is available is to upload the data to a large database with good annotations and metadata, and make a phylogenetic tree of all similar strains. In this way, it might be possible to identify the possible contamination source.

SNPtree, CSIPhylogeny, and NDtree

At CGE, three tools based on Single Nucleotide Polymorphisms (SNPs) are available, which accept both short sequence reads as well as assembled genomes as input. The SNP tools are useful for identifying SNPs in closely related strains. The first developed tool was snpTree [11]. This tool first maps the query genomes to a reference genome selected by the user; either by MUMMER [48] (assembled sequences) or BWA [49] (short sequence reads). The reference genome can either be selected from the CGE database or uploaded by the user. After the mapping, the program localizes SNPs in the genomes by use of SAMtools [50] and nucmer [48]. Following the localization of all SNPs, they are filtered based on user-specified settings. Default settings can also be used, which include a sequencing depth of ten and a minimum distance of ten bases between the SNPs. The SNPs that pass the filtering criteria for each genome are then concatenated to a continuous sequence, and a phylogenetic tree is made based on this multiple alignment. The output files include the alignment in different formats, the tree file, a file showing the individual SNPs in the genomes, VCF[1] files for each genome and a matrix with the difference between the genomes [11].

snpTree is widely used for genome analysis of different species. In 2014 Guio et al. [51] used snpTree to find 218 non-synonymous SNPs in the genome of *Mycobacterium tuberculosis* that could confer resistance towards antibiotics. The service has also been used by Teo et al. [52] to characterize an emerging new pathogen in the hospital environment, *Elizabethkingia anopheles*.

CSIPhylogeny is a further development of snpTree. CSIPhylogeny identifies the SNPs in the same way as snpTree, but is more strict when it comes to filtering (removing) the SNPs. SNPs are filtered if their mapping quality (which is calculated using BWA [49]) is below certain thresholds: If the SNP quality is below 30, or if the sequencing depth is below 10. SNPs are also removed in the filtering step if they are less than ten base pairs from the nearest SNP. Finally a Z-score is calculated for each SNP, which has to be above 1.96, for the SNP to be kept. The Z-score is calculated using the following equation:

$$Z = (X - Y) / \mathrm{sqrt}(X + Y) \tag{5.1}$$

Here X is the number of reads, having the most common nucleotide at that position, and Y the number of reads with any other nucleotide [2, 5]. CSIPhylogeny has the same output files as snpTree.

The main difference between snpTree and CSIPhylogeny lies in the site validation performed by CSIPhylogeny. The validation consists of checking all positions in the analysis and only using those that are considered valid in all the isolates analyzed. The snpTree method performs no validation and assumes all non-SNP positions to be valid, i.e., positions where no SNPs are found or where SNPs has been ignored are assumed to be identical to the base in the reference sequence. This assumption is inconsequently if the isolates compared (including the reference strain) are very closely related. However, the less related the compared isolates are the bigger the consequences will be, because more and more non-SNP positions will not be valid either due to low quality or simply because the DNA in which a SNP is found does not exist in all the genomes of the isolates.

The snpTree server is now deprecated and will no longer be updated. It will remain online, but the CSIPhylogeny server is the recommended tool.

Our third server for constructing phylogenetic trees is NDtree, which constructs the trees based on the number of nucleotide difference found between genomes [5]. It is intentionally made as simple as possible to study which features are important for making accurate phylogenies. NDtree can find a reference genome automati-

[1] VCF (Variant Call Format) files specify a type of text files used for storing sequence variation.

cally using KmerFinder, or may be given a reference sequence by the user. NDtree then maps the query reads to the reference sequence. This is done by splitting the reference sequence and the reads into 17-mers and storing them in a hash table. The 17-mers from the reads are then mapped to the reference to find a match, which is then extended to an optimally scoring ungapped alignment using a match score of 1 and a mismatch score of −3. If the match score is greater than 50 the alignment is used in the SNP calling. A position is significant if the Z-score, as described above in Eq. (5.1), is more than 1.96 and X is ten times larger than Y. If no base can be called based on the above criteria, an "N" is put in the sequence and the position is not used for phylogeny. These called sequences are then compared pairwise, counting the nucleotides that differ. An option can be chosen so all positions called in a given pair of sequences are used, even if that position is not called in one or more of the other sequences. In this case it is advised to increase the Z score cutoff to 3.29. After the pairwise comparisons, the tree is build from the distance matrix, using the UPGMA or neighbor joining (NJ) packages from Phylip. It is recommended to use the UPGMA if the samples are taken at the same time, and otherwise NJ. NDtree output files include the tree file in newick format and the distance matrix.

CSIPhylogeny and NDtree have recently been shown to be more accurate than the older SNPtree method and should be the preferred tools for phylogeny [2].

VTEC Case Study: Determining Phylogeny Using NDtree
The phylogeny of the suspected VTEC isolates was examined using NDtree (see Fig. 5.10). The isolates clustered completely according to serotype, and a clear concordance between serotype and MLST type could be observed. Further, there was a complete agreement with the epidemiological information and the observed clustering [5].

snpTree is available at https://cge.cbs.dtu.dk/services/snpTree.

CSIPhylogeny is available at https://cge.cbs.dtu.dk/services/CSIPhylogeny.

NDtree is available at https://cge.cbs.dtu.dk/services/NDtree.

Metagenomic Samples

Much attention has recently been given to the possibility of diagnosing diseases based on metagenomic samples, since this is faster and simpler than having to initially isolate the bacteria. Hasman et al. [3] were to the best of our knowledge the first to show that metagenomic samples (in this case urine) could be used to diagnose a pathogen without prior knowledge about which species it was. It was found that WGS improved the identification of the cultivated bacteria, and an

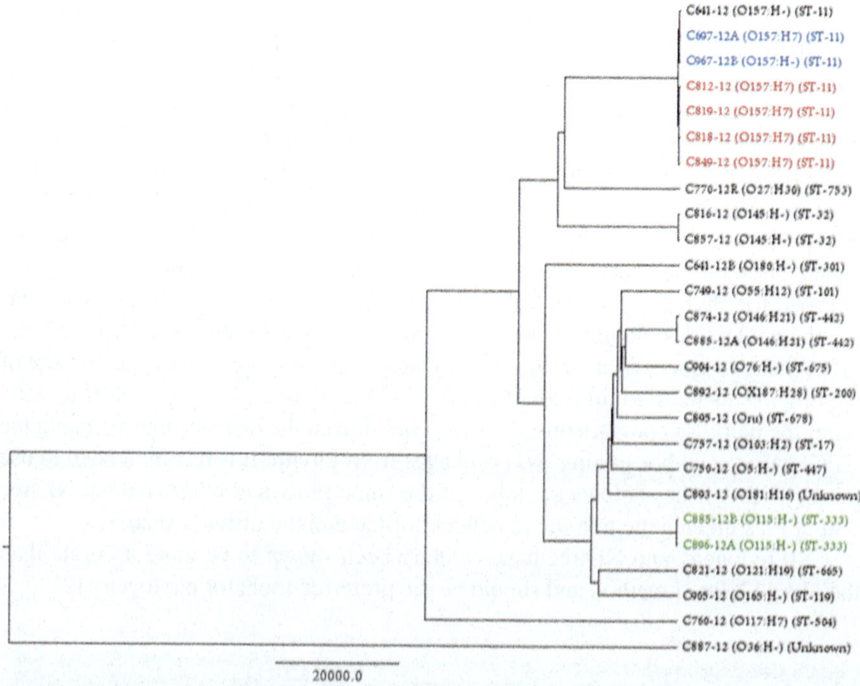

Fig. 5.10 Phylogeny of a subset of the suspected VTEC isolates according to the NDtree method. The tree has been constructed on basis of genome assemblies. Isolates known to be epidemiologically related are shown in the same color, with the red group constituting known outbreak isolates [53]. Serotypes and MLST types are shown for all isolates

almost complete agreement between phenotypic and predicted antimicrobial susceptibilities was observed [3]. Metagenomic analysis could also be used for monitoring purposes and was recently applied to the analysis of toilet waste from 18 international airplanes arriving in Copenhagen, Denmark. It was found that genes encoding antimicrobial resistance were more abundant and also of higher diversity in the planes from South Asia compared to North America. Likewise, the waste from the planes from South Asia indicated a higher presence of *S. enterica* and norovirus. Conversely, the waste from North America contained more *Clostridium difficile* [54].

Work in Progress

Besides finalizing a web-service, which will enable analysis of data from metagenomic samples, several other web-services are in development and expected to be published shortly. These include a method for identification of genes related to restriction-modification systems, pathogenicity islands in S. enterica, and the prediction of the bacterial host of bacteriophages based on the genome sequence of the bacteriophage. To supplement the many stand-alone web-services, we are currently working on a number of improvements that will make the experience even more user-friendly. Specifically, we will include the possibility of batch-uploading several isolates in one go, followed by the automatic execution of several of the analytic tools for typing and phenotyping, and finally a graphical visualization of all isolates on a world map. Furthermore, we will add the option for registered users to manage their sequence data files and keep record of their results. These features are in dire need and their implementation will hopefully make the use of WGS for the analysis of pathogenic microorganisms even faster and more convenient.

Conclusion

Since the start of CGE, a large number of methods have been developed for the purpose of determining the species and subtypes of microorganisms, predict their phenotype, and investigate their phylogeny for epidemiological purposes. The methods have been made publicly available via web-services that are designed to be easy to use/"plug and play"—also for non-bioinformaticians. We will continue to update our existing methods as well as implement new ones, hopefully facilitating that the community can take full advantage of the genomics era.

Conclusion: VTEC Case Study

As a proof-of-concept that WGS-based typing of VTEC could be an attractive alternative in surveillance, real-time WGS-based typing of VTEC was performed during 7 weeks, in parallel to the routine typing carried out at SSI. The study included a set of 46 suspected VTEC isolates that were analyzed using the CGE tools. Overall, the results were concurrent with the routine typing, and the phylogenetic relationship determined was in agreement with epidemiological data. Furthermore, a small ongoing VTEC outbreak [53] was also easily distinguished by the WGS approach. We conclude that WGS-based typing of VTEC is an advantageous alternative to the current routine typing, producing comparable typing results faster and at a similar cost. For complete WGS-based VTEC surveillance WGS O:H serotyping is also needed, which we currently developing.

Acknowledgements The Center for Genomic Epidemiology at the Technical University of Denmark is funded by grant 09-067103/DSF from the Danish Council for Strategic Research.

References

1. Aarestrup FM, Brown EW, Detter C, Gerner-Smidt P, Gilmour D, Harmsen MW, Hendriksen RS, Hewson R, Heymann DL, Johansson K, Ijaz K, Keim PS, Koopmans M, Kroneman A, Lo Fo Wong D, Lund O, Palm D, Sawanpanyalert P, Sobel J, Schlundt J. Integrating genome-based informatics to modernize global disease monitoring, information sharing, and response. Emerg Infect Dis. 2012;18, e1.
2. Kaas RS, Leekitcharoenphon P, Aarestrup FM, Lund O. Solving the problem of comparing whole bacterial genomes across different sequencing platforms. PLoS One. 2014;9, e104984.
3. Hasman H, Saputra D, Sicheritz-Ponten T, Lund O, Svendsen CA, Frimodt-Moller N, Aarestrup FM. Rapid whole-genome sequencing for detection and characterization of microorganisms directly from clinical samples. J Clin Microbiol. 2014;52:139–46.
4. Larsen MV, Cosentino S, Rasmussen S, Friis C, Hasman H, Marvig RL, Jelsbak L, Sicheritz-Ponten T, Ussery DW, Aarestrup FM, Lund O. Multilocus sequence typing of total-genome-sequenced bacteria. J Clin Microbiol. 2012;50:1355–61.
5. Joensen KG, Scheutz F, Lund O, Hasman H, Kaas RS, Nielsen EM, Aarestrup FM. Real-time whole-genome sequencing for routine typing, surveillance, and outbreak detection of verotoxigenic *Escherichia coli*. J Clin Microbiol. 2014;52:1501–10.
6. Lukjancenko O, Thomsen MCF, Larsen MV, Ussery DW. PanFunPro: PAN-genome analysis based on FUNctional PROfiles [v1; ref status: approved with reservations 3]. F1000Research. 2013;2:265. http://f1000r.es/2e1.
7. Cosentino S, Voldby Larsen M, Moller Aarestrup F, Lund O. PathogenFinder—distinguishing friend from foe using bacterial whole genome sequence data. PLoS One. 2013;8, e77302.
8. Carattoli A, Zankari E, Garcia-Fernandez A, Voldby Larsen M, Lund O, Villa L, Moller Aarestrup F, Hasman H. In silico detection and typing of plasmids using PlasmidFinder and plasmid multilocus sequence typing. Antimicrob Agents Chemother. 2014;58:3895–903.
9. Zankari E, Hasman H, Cosentino S, Vestergaard M, Rasmussen S, Lund O, Aarestrup FM, Larsen MV. Identification of acquired antimicrobial resistance genes. J Antimicrob Chemother. 2012;67:2640–4.
10. Joensen KG, Tetzschner AM, Iguchi A, Aarestrup FM, Scheutz F. Rapid and easy in silico serotyping of *Escherichia coli* isolates by use of whole-genome sequencing data. J Clin Microbiol. 2015;53:2410–26.
11. Leekitcharoenphon P, Mortensen RK, Thomsen MCF, Friis C, Rasmussen S, Aarestrup FM. snpTree—a web-server to identify and construct SNP trees from whole genome sequence data. BMC Genomics. 2012;13 Suppl 7:S6.
12. Larsen J, Enright MC, Godoy D, Spratt BG, Larsen AR, Skov RL. Multilocus sequence typing scheme for *Staphylococcus aureus*: revision of the gmk locus. J Clin Microbiol. 2012;50:2538–9.
13. Boerlin P, McEwen SA, Boerlin-Petzold F, Wilson JB, Johnson RP, Gyles CL. Associations between virulence factors of Shiga toxin-producing *Escherichia coli* and disease in humans. J Clin Microbiol. 1999;37:497–503.
14. Karmali MA. Infection by verocytotoxin-producing *Escherichia coli*. Clin Microbiol Rev. 1989;2:15–38.
15. Karch H, Tarr PI, Bielaszewska M. Enterohaemorrhagic *Escherichia coli* in human medicine. Int J Med Microbiol. 2005;295:405–18.
16. CDC. Centers for Disease Control and Prevention—*E. coli*. General Information. 2012. http://www.cdc.gov/ecoli/general/#complications.
17. Richter M, Rossello-Mora R. Shifting the genomic gold standard for the prokaryotic species definition. Proc Natl Acad Sci U S A. 2009;106:19126–31.
18. Fox GE, Peckman KJ, Woese CE. Comparative cataloging of 16S ribosomal ribonucleic acid: molecular approach to procaryotic systematics. Int J Syst Bacteriol. 1977;27:44–57.
19. DeSantis TZ, Hugenholtz P, Larsen N, Rojas M, Brodie EL, Keller K, Huber T, Dalevi D, Hu P, Andersen GL. Greengenes, a chimera-checked 16S rRNA gene database and workbench compatible with ARB. Appl Environ Microbiol. 2006;72:5069–72.

20. Pruesse E, Quast C, Knittel K, Fuchs BM, Ludwig W, Peplies J, Glockner FO. SILVA: a comprehensive online resource for quality checked and aligned ribosomal RNA sequence data compatible with ARB. Nucleic Acids Res. 2007;35:7188–96.
21. Kampfer P. Systematics of prokaryotes: the state of the art. Antonie Van Leeuwenhoek. 2012;101:3–11.
22. Tindall BJ, Rossello-Mora R, Busse HJ, Ludwig W, Kampfer P. Notes on the characterization of prokaryote strains for taxonomic purposes. Int J Syst Evol Microbiol. 2010;60:249–66.
23. Tindall BJ, Schneider S, Lapidus A, Copeland A, Glavina Del Rio T, Nolan M, Lucas S, Chen F, Tice H, Cheng JF, Saunders E, Bruce D, Goodwin L, Pitluck S, Mikhailova N, Pati A, Ivanova N, Mavrommatis K, Chen A, Palaniappan K, Chain P, Land M, Hauser L, Chang YJ, Jeffries CD, Brettin T, Han C, Rohde M, Goker M, Bristow J, Eisen JA, Markowitz V, Hugenholtz P, Klenk HP, Kyrpides NC, Detter JC. Complete genome sequence of *Halomicrobium mukohataei* type strain (arg-2). Stand Genomic Sci. 2009;1:270–7.
24. Klenk HP, Goker M. En route to a genome-based classification of Archaea and Bacteria? Syst Appl Microbiol. 2010;33:175–82.
25. Larsen MV, Cosentino S, Lukjancenko O, Saputra D, Rasmussen S, Hasman H, Sicheritz-Ponten T, Aarestrup FM, Ussery DW, Lund O. Benchmarking of methods for genomic taxonomy. J Clin Microbiol. 2014;52(5):1529–39.
26. Lukjancenko O, Ussery DW. Vibrio chromosome-specific families. Front Microbiol. 2014;5:73.
27. Goujon M, McWilliam H, Li W, Valentin F, Squizzato S, Paern J, Lopez R. A new bioinformatics analysis tools framework at EMBL-EBI. Nucleic Acids Res. 2010;38:W695–9.
28. Woese CR, Fox GE. Phylogenetic structure of the prokaryotic domain: the primary kingdoms. Proc Natl Acad Sci U S A. 1977;74:5088–90.
29. Helgason E, Okstad OA, Caugant DA, Johansen HA, Fouet A, Mock M, Hegna I, Kolsto AB. *Bacillus anthracis*, *Bacillus cereus*, and *Bacillus thuringiensis*—one species on the basis of genetic evidence. Appl Environ Microbiol. 2000;66:2627–30.
30. Karaolis DK, Lan R, Reeves PR. Sequence variation in *Shigella sonnei* (Sonnei), a pathogenic clone of *Escherichia coli*, over four continents and 41 years. J Clin Microbiol. 1994;32:796–802.
31. Lan R, Reeves PR. *Escherichia coli* in disguise: molecular origins of *Shigella*. Microbes Infect. 2002;4:1125–32.
32. Maiden MC, Bygraves JA, Feil E, Morelli G, Russell JE, Urwin R, Zhang Q, Zhou J, Zurth K, Caugant DA, Feavers IM, Achtman M, Spratt BG. Multilocus sequence typing: a portable approach to the identification of clones within populations of pathogenic microorganisms. Proc Natl Acad Sci U S A. 1998;95:3140–5.
33. Jaureguy F, Landraud L, Passet V, Diancourt L, Frapy E, Guigon G, Carbonnelle E, Lortholary O, Clermont O, Denamur E, Picard B, Nassif X, Brisse S. Phylogenetic and genomic diversity of human bacteremic *Escherichia coli* strains. BMC Genomics. 2008;9:560.
34. Wirth T, Falush D, Lan R, Colles F, Mensa P, Wieler LH, Karch H, Reeves PR, Maiden MC, Ochman H, Achtman M. Sex and virulence in *Escherichia coli*: an evolutionary perspective. Mol Microbiol. 2006;60:1136–51.
35. Griffiths D, Fawley W, Kachrimanidou M, Bowden R, Crook DW, Fung R, Golubchik T, Harding RM, Jeffery KJ, Jolley KA, Kirton R, Peto TE, Rees G, Stoesser N, Vaughan A, Walker AS, Young BC, Wilcox M, Dingle KE. Multilocus sequence typing of *Clostridium difficile*. J Clin Microbiol. 2010;48:770–8.
36. Lemee L, Dhalluin A, Pestel-Caron M, Lemeland JF, Pons JL. Multilocus sequence typing analysis of human and animal *Clostridium difficile* isolates of various toxigenic types. J Clin Microbiol. 2004;42:2609–17.
37. Altschul SF, Madden TL, Schaffer AA, Zhang J, Zhang Z, Miller W, Lipman DJ. Gapped BLAST and PSI-BLAST: a new generation of protein database search programs. Nucleic Acids Res. 1997;25:3389–402.
38. Hendriksen RS, Joensen KG, Lukwesa-Musyani C, Kalondaa A, Leekitcharoenphon P, Nakazwe R, Aarestrup FM, Hasman H, Mwansa JC. Extremely drug-resistant *Salmonella enterica* serovar Senftenberg infections in patients in Zambia. J Clin Microbiol. 2013;51:284–6.

39. Rodriguez-Rivera LD, Moreno Switt AI, Degoricija L, Fang R, Cummings CA, Furtado MR, Wiedmann M, den Bakker HC. Genomic characterization of *Salmonella* Cerro ST367, an emerging *Salmonella* subtype in cattle in the United States. BMC Genomics. 2014;15:427.
40. Stegger M, Aziz M, Chroboczek T, Price LB, Ronco T, Kiil K, Skov RL, Laurent F, Andersen PS. Genome analysis of *Staphylococcus aureus* ST291, a double locus variant of ST398, reveals a distinct genetic lineage. PLoS One. 2013;8, e63008.
41. Zhang S, Yin Y, Jones MB, Zhang Z, Deatherage Kaiser BL, Dinsmore BA, Fitzgerald C, Fields PI, Deng X. Salmonella serotype determination utilizing high-throughput genome sequencing data. J Clin Microbiol. 2015;53:1685–92.
42. Carattoli A, Bertini A, Villa L, Falbo V, Hopkins KL, Threlfall EJ. Identification of plasmids by PCR-based replicon typing. J Microbiol Methods. 2005;63:219–28.
43. Zankari E, Hasman H, Kaas RS, Seyfarth AM, Agerso Y, Lund O, Larsen MV, Aarestrup FM. Genotyping using whole-genome sequencing is a realistic alternative to surveillance based on phenotypic antimicrobial susceptibility testing. J Antimicrob Chemother. 2013;68:771–7.
44. Villa L, Feudi C, Fortini D, Garcia-Fernandez A, Carattoli A. Genomics of KPC-producing *Klebsiella pneumoniae* sequence type 512 clone highlights the role of RamR and ribosomal S10 protein mutations in conferring tigecycline resistance. Antimicrob Agents Chemother. 2014;58:1707–12.
45. Leonard SR, Lacher DW, Elkins CA, Jung CM. Draft genome sequence of the multidrug-resistant *Escherichia coli* strain LR09, isolated from a wastewater treatment plant. Genome Announc. 2014;2, e00272-14.
46. Kroft BS, Brown EW, Meng J, Gonzalez-Escalona N. Draft genome sequences of two Salmonella strains from the SARA collection, SARA64 (Muenchen) and SARA33 (Heidelberg). Provide insight into their antibiotic resistance. Genome Announc. 2013;1, e00806-13.
47. Andreatta M, Nielsen M, Moller Aarestrup F, Lund O. In silico prediction of human pathogenicity in the gamma-proteobacteria. PLoS One. 2010;5, e13680.
48. Delcher AL, Phillippy A, Carlton J, Salzberg SL. Fast algorithms for large-scale genome alignment and comparison. Nucleic Acids Res. 2002;30:2478–83.
49. Li H, Durbin R. Fast and accurate short read alignment with Burrows-Wheeler transform. Bioinformatics. 2009;25:1754–60.
50. Li H, Handsaker B, Wysoker A, Fennell T, Ruan J, Homer N, Marth G, Abecasis G, Durbin R. The sequence alignment/map format and SAM tools. Bioinformatics. 2009;25:2078–9.
51. Guio H, Tarazona D, Galarza M, Borda V, Curitomay R. Genome analysis of 17 extensively drug-resistant strains reveals new potential mutations for resistance. Genome Announc. 2014;2, e00759-14.
52. Teo J, Tan SY, Liu Y, Tay M, Ding Y, Li Y, Kjelleberg S, Givskov M, Lin RT, Yang L. Comparative genomic analysis of malaria mosquito vector-associated novel pathogen *Elizabethkingia anophelis*. Genome Biol Evol. 2014;6:1158–65.
53. Soborg B, Lassen SG, Muller L, Jensen T, Ethelberg S, Molbak K, Scheutz F. A verocytotoxin-producing *E. coli* outbreak with a surprisingly high risk of haemolytic uraemic syndrome, Denmark, September–October 2012. Euro Surveill. 2013;18:1–3.
54. Nordahl Petersen T, Rasmussen S, Hasman H, Caroe C, Baelum J, Schultz AC, Bergmark L, Svendsen CA, Lund O, Sicheritz-Ponten T, Aarestrup FM. Meta-genomic analysis of toilet waste from long distance flights; a step towards global surveillance of infectious diseases and antimicrobial resistance. Sci Rep. 2015;5:11444.

Chapter 6
Genomic Diversity in *Salmonella enterica*

John Wain and Justin O'Grady

Introduction

There are two species within the genus *Salmonella* (*bongori* and *enterica)* that occupy different ecological niches and are genomically distinct—*S. bongori* has evolved to exploit cold blooded animals, as a host, through the acquisition of genes encoding specialised effector proteins secreted through the same *Salmonella* secretion systems as *S. enterica* [1]; *S. bongori* is not well studied. On the other hand there is a wealth of knowledge about the biology of *S.* enterica, an important human and animal pathogen, which is now being translated into new methods for diagnosing individuals and tracking outbreaks. There have been many books and chapters on *Salmonella* over the past decade but the field is moving fast and this chapter will focus on new technology and the translation of that technology into clinical practice and public health benefit.

Genetic Variation Within *Salmonella enterica*

Salmonella enterica is split into sub-species as shown in Fig. 6.1. The six subspecies can be differentiated by many tests including their genetic content [2] and multi locus sequence typing [3]. The *S. enterica* sub-species *enterica* (also known as subspecies I) has been studied in much more detail because strains from this sub-species cause most of the *Salmonella* infection in man and farmed animals; it

J. Wain (✉) • J. O'Grady
Norwich Medical School, University of East Anglia, Room 2.28, Bob Champion Research and Education Building, James Watson Road, Norwich Research Park, Norwich NR4 7UQ, UK
e-mail: j.wain@uea.ac.uk; Justin.OGrady@uea.ac.uk

X. Deng et al. (eds.), *Applied Genomics of Foodborne Pathogens*, Food Microbiology and Food Safety, DOI 10.1007/978-3-319-43751-4_6

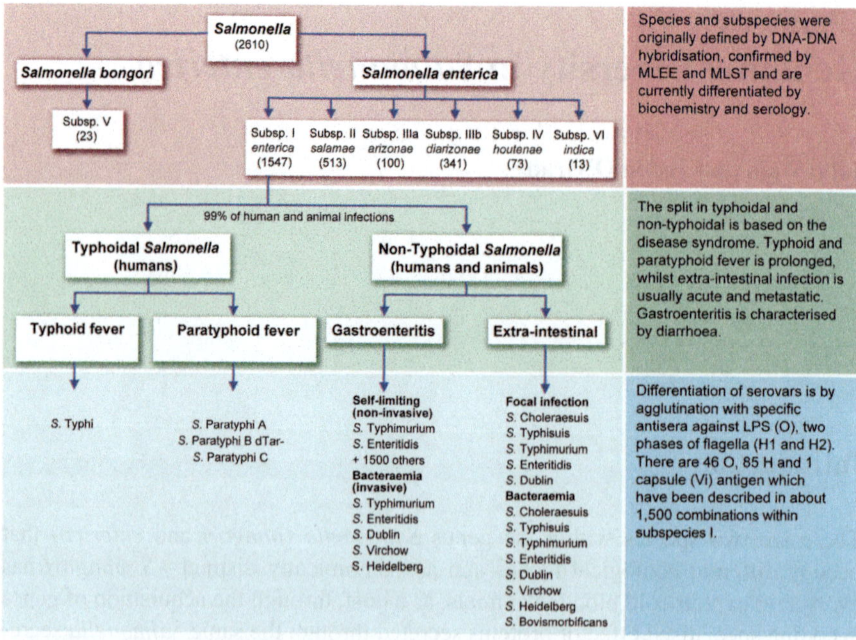

Fig. 6.1 Overview of the current classification of *Salmonella enterica*

contains over 1500 serotypes (also called serovars if formally accepted in standard nomenclature). Strains from within a single serotype show different levels of variation. Some (monomorphic) serotypes contain a single strain in a single lineage most often adapted to a single host (e.g., *S.* Typhi) whilst other (polymorphic) serotypes contain multiple lineages of strains which may or may not be host adapted (e.g., *S.* Choleraesuis). The level of genomic variation present within a serotype can be driven by the biology of the serotype; recent acquisition by *S.* Agona of indels (51 bacteriophages, ten plasmids, and six integrative conjugational elements) has led to high diversity according to pulsed-field gel electrophoresis (PFGE) [4] but resulting in inaccurate clustering by PFGE—there is no evidence of selection and so we must assume that this variation is transient and will not become fixed in the population. The level of genomic diversity may also be because of the methods used for defining the serotype. Serotyping depends on antibody based detection of surface antigens and antibody preparations against different antigens have different levels of specificity depending on the strains available for removing cross reacting antibodies. This non-standard level of selectivity within each antibody preparation naturally leads to non-standard variation within each serotype and is a problem when assessing new methods for typing and tracking if traditional serotyping is used as a gold standard. However, we need to name *Salmonella*. The cause of a bacterial infection needs to be identified so that we can correctly diagnose the infection the next time the bacterium is isolated and also for risk assessment when bacteria are identified from food. It is not realistic to expect to define the level of genomic variation expected within a group of bacteria which have been given a name (e.g., a serotype) because groups of bacteria with the same name do not all cause the same disease nor are they all adapted to the same host in the same way—i.e., niche adapted. The genetic variation present in niche adapted bacteria is intrinsically variable because; different niches provide access to different levels of diversity; different bacteria became host adapted at different time points; the population supported by different niches varies in size and there is different selection pressures in different niches. Even in the absence of selection genetic diversity is a function of time multiplied by population size; therefore it is naïve to expect that the genetic variation within each group we are trying to name will be constant. For outbreak investigation the situation is different, we need to sub-type, at a more granular level, to distinguish between the last common source of two strains so that we can tell if they are from the outbreak under investigation or not. It is, again, impossible to define a level of variation which can define the strains as outbreak or non-outbreak from all outbreaks because, although for *Salmonella* a point source (single strain) is often the cause, outbreaks have different characteristics; size of outbreak; and how long the outbreak continues. It is with this background in mind that we need to look at how to develop, and interpret the data on genetic variation in *Salmonella*.

Host Adaptation and Host Restriction vs Identification and Typing

Salmonella evolve through a process of diversification from a point source followed by colonisation of a new niche by a strain capable of doing so [5]. There are two likely events which result in the exploitation of a new niche. Firstly, chance—a serotype which comes into contact with a new host is already capable of colonising it. If this is a rare event then only one strain will pass through this bottleneck and most likely become isolated from other members. In time, if the strain adapts through alteration of O and H antigens, it will be recognised as a new serotype and the whole population of that serotype will be clonally derived from a single ancestor. If many strains can colonise the new host a variety of strains go through the bottleneck and adapt through convergent evolution to the most successful type capable of infection in the population of the new host. The second possibility is that it is the acquisition of a new trait (virulence factor) allows a specific strain of the serotype to colonise the new host. In this case, again, the bottleneck allows divergence of the strain away from the other members of the serotype. Superimposed onto this background is the possibility of convergent evolution, where two different organisms colonise the same niche (e.g., *S.* Typhi and *S.* Paratyphi A) [6]. Superimposed on the purifying selection of adaptation to niche is also a diversifying selection on the antigens. Host immune responses are aimed at the O and H antigens and so changes in these antigens can allow the pathogen to escape destruction by the host. This can occur through regulated changes in phase, through changes in the amino acid structure of each antigen—seen particularly in flagella genes—but may be the result of horizontal exchange of flagella genes which leads to the same flagella antigen occurring in non-related *Salmonella* backgrounds.

Salmonella enterica Subspecies *Enterica* Serotype Enteritidis, an Example of a Monophyletic Group of Strains

S. Enteritidis is the most common *Salmonella* serotype isolated from human disease globally [7] and is responsible for approximately 60 % of *Salmonella* infections in humans, making it the leading cause of salmonellosis in Europe [8]. The *S.* Enteritidis population, as with other monomorphic bacterial populations [9] contains insufficient variation for differentiation of strains using traditional typing methods so that improved subtyping of isolates is needed to support classical epidemiological data for the detection of outbreaks and identification of the vehicle of infection [10]. The Kaufman-White scheme defines *S.* Enteritidis by the presence of the O9 somatic antigen and a single flagella antigen, phase 1, g.m. The serotype can be isolated from many animal species including cows [11], pigs [12], guinea pigs [13] and even hedgehogs [14]. Molecular analysis reveals non-host adapted adhesion factors [15] and yet the overwhelming majority of infections seen in humans is from chickens [16], suggesting that at some level *S.* Enteritidis is adapted to birds. Sub-typing using traditional methods has always been problematic for *S.* Enteritidis [10, 17] and plasmid profiling is often used to define an outbreak strain [18]. Phage typing and pulsed-field profiles of fragmented chromosomal DNA are not useful for outbreak investigation [19] and multi-locus variable number tandem repeat (VNTR) typing (MLVA), whilst useful, does not allow the identification of all outbreaks [20]. Multi-locus sequencing typing (MLST) of seven housekeeping genes reveals several sequence types (STs) within the group the *S.* Enteritidis group; one dominant ST (ST11) is linked through an alteration in one allele only to several single locus variants [21]—this is good for serotype definition (all singe locus variants from STs existing in the database can be identified as *S.* Enteritidis) but is not good for outbreak investigation (the majority of isolates, whether epidemiologically associated or not, have the same ST). Only whole genome sequencing [22] shows promise for improving typing. For a century *S.* Enteritidis has been considered as closely related to (other bird adapted, serotypes) *S.* Pullorum and *S.* Gallinarum [23] and sequence based studies of host adaption for this group [24] have confirmed the relationship. Detailed studies looking at genomic diversity suggest that the *S.* Enteritidis population consists of two clades [25]: a classical clade and a second clade; S Gallinarum and S, Pullorum have most likely evolved from the second clade. Granular typing at this resolution, together with host colonisation studies, should eventually lead to resolution of the link between bacterial genetics and host colonisation and adaptation—understanding this is key to being able to interpret the fine typing data generated by whole genome sequencing.

There are of course, even for "good" serotypes always occasional strains which have the same antigenic profile but are entirely unrelated but this is easily detected by MLST [21].

The serotype *S*. Enteritidis is a clonally derived, monophyletic, group of bacterial strains showing exquisite host adaptation. Therefore *S*. Enteritidis is an easily defined serotype that represents a group of strains that all cause a similar infection and studies of evolution and host adaptation are straight forward. However, it is clear that for subtyping during outbreak investigation of *S*. Enteritidis, we need to detect recently acquired genetic diversity such as plasmid carriage or single nucleotide polymorphisms (SNPs) and that this is best achieved by genome sequencing [26].

Salmonella enterica Subspecies *Enterica* Serotype Montevideo, an Example of a Polyphyletic Group of Strains

Isolations of *S*. Montevideo rank amongst the top 10 serotypes of *Salmonella* isolated from food borne outbreaks in both Europe and the USA. The serotype is split by MLST into at least three sequence types which, although related, do not share alleles and so this serotype does not conform to the definition of monomorphic [21]. Outbreaks however, are difficult to investigate because of the lack of variation in circulating strains, which are usually from the same MLST group. As with *S*. Enteritidis, *S*. Montevideo has a reputation, particularly the USA, as being difficult to sub-type. Traditional methods, such a PFGE, fail to distinguish between isolates and alternatives are actively being sought. Whole genome sequencing is, therefore, an obvious choice for the investigation of outbreaks of *S*. Montevideo. The successful investigation of an outbreak caused by isolates with a single PFGE pattern [27] illustrates the utility of whole genome sequencing for this serotype. Further analysis of the population structure of *S*. Montevideo by whole genome sequencing [28] reveals that the variation between the previously described sub-groups of Montevideo is driven by large chromosomal deletion/insertion events of bacteriophage and plasmid associated sequence and, perhaps more significantly, the variation within each cluster is mainly through SNPs. Fortunately for the investigation of outbreaks caused by this serotype there is enough variation, even in the core genome, to differentiate outbreak strains from non-outbreak strains within a ST. This work [28] was very thorough and compared results from previous genome sequencing studies to show that if the analysis is standardised through protocol control then the resulting phylogeny, and so the epidemiological investigation, is independent of the sequencing method. They were able to "…effectively delineate the scope of the outbreak." This is an important finding as it demonstrates that whole genome sequence typing can be used, at least in *Salmonella* outbreaks, as part of the outbreak definition. Whilst this may be true for the monomorphic serotypes of *Salmonella enterica* more needs to be done to understand the less well delineated serotypes.

Salmonella enterica Subspecies *Enterica* Serotype Typhimurium, a "Challenging" Group of Strains

Strains of serotype *S.* Typhimurium show much more diversity than strains of *S.* Enteritidis (personal experience). The population structure, by MLST, has a few more sequence types in the major cluster than for *S.* Enteritidis (28 vs 21) but there is nothing obvious from the MLST data which would explain the difference in variation experienced by microbiologists. What is clear is that some of the sequence types linked to the major cluster actually have a different phenotype. For example, sequence type ST128 is a single locus variant of the most common sequence type, ST19, but consists of pigeon adapted *S.* Typhimurium strains, also defined by phage typing as DT2. It seems that *S.* Typhimurium, although a very well defined group, conceals biological differences within its ancestry which may not have had time to emerge as new sequence type clusters or serotypes. This is also reflected in the usefulness of phage typing and MLVA when investigating outbreaks; both have proven to be very useful tools [17, 29]. However, this situation can change. For example, the emergence of a dominant phage type (DT170, accounting for around 40 % of S. Typhimurium isolates) [30] threatens the utility of phage typing and MLVA in Australia; the spread of ST313 across sub-Saharan Africa threatens our ability to track local outbreaks and trace the source of contamination in the food chain [31]; and the emergence of monophasic strains in pigs (with no expression of flagella proteins of the second phase) threaten our ability to give strains of *S.* Typhimurium the correct serotype designation [32].

So, our current capability to track, trace and investigate *S.* Typhimurium is sufficient but may not be so for much longer. In Australia, an investigation of *S.* Typhimurium DT170 using whole genome sequencing revealed that the method has strong utility in outbreak investigation [33] but also raised questions on how to decide the cut-off necessary for a rule-out decision: not outbreak associated. Isolates of *S.* Typhimurium from within each of five different outbreak investigations differed, on average, by three or four single nucleotide polymorphisms (SNPs) in the core genome (underlining the importance of sequencing accuracy) whereas between outbreaks there was 10–20 SNPs. Whilst this is useful for epidemiology, new ways of analysing the differences between genomes which do not rely on the use of genes shared by all members of the group under investigation are required. In Australia, the current population structure of *S.* Typhimurium is the result of "right time right place" events which allow strains to expand clonally without advantage or selection. The genetic diversity seen is, therefore, the result of drift and so represents the clock rate for mutations occurring in the background of the strains.

The investigation of *S.* Typhimurium sequence type ST313 isolates from sub-Saharan Africa however, tells a very different story [34]. Two closely related lineages of *S.* Typhimurium emerged but the second lineage replaced the first in under 10 years. We believe this was the result of the acquisition of plasmid borne genes. Within single countries the usual pattern was for *S.* Typhimurium to occur as a discrete event, followed by spread within the country. For some countries the strain was introduced more than once. As the strain spreads more widely we would expect more commonly to find multiple strains co-located geographically. Independent acquisition of a Tn21 element encoding MDR genes into different sites of the resident plasmids may have facilitated transmission but the later acquisition of a plasmid borne chloramphenicol resistance gene by a lineage II strain resulted in the clonal replacement of lineage I. This probably occurred between 2003 and 2005 and the link is supported by phylogenetic analysis of plasmids from ST313 strains which matched the phylogeny of core genes on the chromosome. Such powerful selection can result in phenotypes being generated by single genetic events without there being any impact on the phylogenetic signal. An association between acquisition of plasmids and increased transmission has also been observed in *S.* Typhi [35]. Acquisition of an antibiotic resistance plasmid often converts a susceptible strain to a resistant strain—but the name of a group of bacteria does not change because they have become resistant to antibiotics, however, if the difference is host range then the situation is less clear. If *S.* Typhimurium ST313 has, as some suggest, become adapted to the human host and transmission is mainly human to human then this conflicts with our current understanding that *S.* Typhimurium are animal pathogens which transmit to humans as zoonotic infections. The importance of understanding the biology of a serotype is illustrated by the investigation of the monophasic *S.* Typhimurium DT191a [36] which was focused on mice even though the main risk factor from trawling questionnaires was ownership of reptiles. The decision to investigate mice was taken because prior knowledge of the biology of *S.* Typhimurium suggested reptiles were unlikely to be the primary source of this *Salmonella* serotype and led to the tracking of the infection source to feeder mice imported into the UK.

Another apparent link between genetic structure and virulence in *S.* Typhimurium is the emergence of monophasic isolates [29, 36–39]. Reference laboratories across the world name strains using the Kaufman and White Scheme which defines *S.* Typhimurium by the antigenic structure 4:i:1,2. Several laboratories reported the emergence of isolates which were 4:i:-, that is they had no second phase for the flagella antigen. Investigation of the *fljB* gene [40, 41], which encodes the phase II antigen, revealed multiple genetic mechanisms for this phenotype and the possible association of a new genomic island. Phage types were mostly typical of *S.* Typhimurium and MLST clustered the strains amongst typical *S.* Typhimurium STs. A survey of isolates in the UK [29] concluded: "Monophasic variants of serovar Typhimurium have already caused substantial outbreaks in several countries and continue to pose a public health risk. Reliable detection of monophasic variants of serovar Typhimurium is important to ascertain the impact the emergence of these strains is having on the food chain and the number of human infections, and to monitor control efforts. Legislation of the EU has now been redrafted to include serovar 4:i:- in actions taken to detect and control *Salmonella* serovars of public health significance in laying hens (Commission Regulation (EU) No. 517/2011). In order to more accurately identify these isolates, the UK reference laboratory has been determining the full antigenic structure of all presumptive O:4 isolates since the beginning of 2012 in addition to performing phage typing for identification of serovar Typhimurium and its variants. At present molecular methods will not be applied due to the large number of isolates received in the laboratory, but new sequence based methods for identification of serovar Typhimurium and its variants will be assessed as they become available."

Whole genome sequencing is now, in 2015, being used to assess these strains in the UK. Several points arise from all of the work on monophasic *S.* Typhimurium. It is clear that these isolates are derived from different lineages within the *S.* Typhimurium population and the genetic event giving rise to each monophasic phenotype are different, therefore these organisms do not constitute a recent ancestral group and so cannot be differentiated from other *S.* Typhimurium phylogenetically. Furthermore, monophasic isolates have been isolated from a range of hosts and so do not seem to show, as a group, any specific host adaptation. However, large scale genomic investigation into monophasic *S.* Typhimurium is currently underway in the UK and early analysis would suggest that there is granularity within the group and that there may be selection for growth in certain hosts. If this turns out to be true then a new sub-group of host adapted *S.* Typhimurium could emerge to take its place alongside DT2 and ST313.

Salmonella enterica Subspecies *Enterica* Serotype Choleraesuis, an Example of a Polyphyletic Group of Strains with Different Host Adaptations

The serotype 6/7:c:1,5 has been split into two biotypes using the phenotypic tests dulcitol and tartrate: *S.* Paratyphi C (associated with a form of enteric fever in humans) and *S.* Choleraesuis (septicemia in swine and immunocompromised humans). There is a very rare third biotype, Typhisuis (chronic paratyphoid/caseous lymphadenitis in swine), which is largely ignored in clinical and veterinary medicine. Some *S.* Paratyphi C isolates express the Vi capsular antigen and so can be recognised directly but further biotyping using H_2S production and mucate utilisation are used to subdivide *S.* Choleraesuis into: sensu stricto, Kunzendorf and Decatur. Using MLST [21] *S.* Paratyphi C is found as a distinct sequence type cluster but is closely related to both *S.* Typhisuis, and Choleraesuis. Even the distinction between var Kunzendorf and var sensu strictu within Choleraesuis can be recognised by MLST, suggesting a different ancestry for these biotypes. However, *S.* Choleraesuis var. Decatur is very variable and has no common ancestry; this biotype consists of isolates from at least seven unrelated sequence types. The assignment of isolates of *S.* Decatur, *S.* Paratyphi C and *S.* Choleraesuis to the same serotype is, therefore, misleading. Sequencing has helped to explain why such different strains have the same antigenic formula; the nucleotide sequence of the *fliC* allele encoding Hc is variable (there are 12 amino acid differences in the Hc *fliC* allele). Although matched, statistically meaningful, comparisons with other *fliC* alleles have not been carried out, there would seem to be much more variation than seen in the Hd *fliC* allele for *S.* Typhi. Whole genome sequencing has not been reported for this group at the date of publication but it is clear that the routine use of the technology would easily reclassify the 6/7:c:1,5 organisms into more useful groups with biological significance. Choleraesuis is now a rare serotype in Europe and America and so sub-typing is not an issue. Outbreak investigation is made easier when rare strains are identified—it is clear that the strain is part of the outbreak, because there are no non-outbreak strains of the same serotype circulating. However, the main concern with *S.* Choleraesuis is tracking the transmission of isolates through the food chain. Although there is enough genetic diversity for this to be a straight forward task we do not yet know where to put the cut-off in diversity between strains that are part of a transmission chain and those that are not. The Choleraesuis group of organisms contains one large cluster of strains as defined by MLST which includes var sensu strictu and var Kunzendorf. These are all linked in a network though single locus variants and the genomic diversity has not yet been defined—this work still needs to be done for this potentially fatal pathogen.

The Special Case of *Salmonella* that Cause Human Enteric Fever

There are only two serotypes of *Salmonella* that reliably cause enteric fever in humans: *S.* Typhi, and *S.* Paratyphi A. The other paratyphoid serotypes (*S.* Paratyphi B and C) are not well defined as serotypes and cause a disease clinically distinguishable from classic enteric fever. In this section, therefore, we shall focus on *S.* Typhi and *S.* Paratyphi A. Both can be food borne pathogens and for both the population structure is monomorphic.

The emergence of *S.* Paratyphi A occurred around 450 years ago and its adaptation to the human host is ongoing and is mainly through recent events [42] suggesting that host adapted *Salmonella* serotypes arise through chance environmental events, including geographical spread and/or transmissions to naïve hosts and then adaption and or mutation. Population genome data suggests that *S.* Typhi is older than *S.* Paratyphi A and the event which allowed the ancestor of Paratyphi A to colonise humans may have been the acquisition of a significant portion of the *S.* Typhi genome [43]. Since that event, local variation arose with only transient advantage being gained and the loss of most mutations through purifying selection—which removes the disadvantageous mutations through competition and cost. Recent changes in the genomes of *S.* Paratyphi A include multiple mutations and acquisitions or losses of genes including bacteriophages, genomic islands and plasmids. For *S.* Paratyphi A there are 1560 acquired genes in the accessory genome: plasmid genes are found as 11 separate gene clusters; bacteriophage genes are present in 23 regions of the chromosome; and there are two gene clusters integrated into the chromosome [42]. Gain, and survival, of the same genetic change in different strains is indicative of selection, suggesting that the cargo genes (non-viral genes associated with bacteriophage sequence; or non-plasmid genes associated with plasmid sequence) have changed the bacterial phenotype in a way which resulted in positive selection. It is noteworthy that only the plasmid genes show this tendency. Furthermore, there are no known virulence genes described and the variation is seen mainly in the tips of the phylogenetic tree (suggestion recent acquisition) which may be pruned if not under positive selection.

For *S.* Typhi, the emergence of an epidemic clone may change this situation. *S.* Typhi is a monomorphic serotype which emerged at least 50,000 years ago [44] for which the last common ancestor remains extant [45]. This means there has been no genetic sweep since *S.* Typhi emerged and that all strains have survived and spread across the globe. Diversity was, presumably, driven initially by the colonisation of humans by a common ancestor, isolation from other *Salmonella* in a host adapted niche, and expansion in population size. Host restriction of *S.* Typhi to humans came later, driven by the lack of purifying selection, in the absence of competition, and accumulated gene loss mainly through pseudogene formation. The impact of antibiotic use, however, may be changing this picture. Antibiotic resistance emerged almost immediately after the introduction of chloramphenicol for the treatment of typhoid fever [46]. These early chloramphenicol resistant strains had an associated cost—in the 1970s treatment of patients infected with chloramphenicol resistant *S.* Typhi responded more quickly to trimethoprim therapy than did the patients infected with susceptible strains [47]. In the 1990s, however, the situation changed; in Pakistan, patients with multi-drug resistance plasmids where harder to treat with cephalosporin antibiotics than plasmid free strains, even though both types of isolate were equally susceptible [48]. The plasmid containing strains also gave higher counts in the blood than plasmid negative strains suggesting that, unlike S. Paratyphi A, a strain with increased virulence had emerged [49]. Further work identified a new clone of *S.* Typhi, haplotype; H58, spreading globally, which had very little variation at the genome level [50]. This has practical implications for public health. The reservoir for *S.* Typhi in the human population is carriers and sequencing of isolates from carriers (not published) and a case for which there has been direct transmission shows only 0–3 SNPs in the 4 megabases or so of core sequence. As the H58 clone spreads, it seems inevitable that the genetic variation in the population of *S.* Typhi as a whole will decrease. The recent use of long range sequencing with nanopore technology [51] to identify the presence, location and structure of a genomic island may give us a way forward. The *S.* Typhi story is a warning, the variation in core genes between *S.* Typhi H58 strains may not be sufficient for epidemiological investigation of outbreaks nor for the tracking of food contamination globally. This major bacterial pathogen, once easily characterised by low resolution methods such as PFGE, could become intractable, through the global sweep of a single clone.

Detecting the Variation in *Salmonella*

The variation between *Salmonella* serotypes detectable by antigen recognition is also definable by straight forward genome sample sequencing techniques. Sample sequencing involves the sequencing of selected regions of the bacterial genome. If the regions selected are conserved then discrimination is low but the clustering is robust and related to ancestry (e.g. MLST). On the other hand, if the region selected is highly variable then a higher index of discrimination is possible but the relationship between isolates is not as clear. This is the basis of virulence typing, where the genotype may predict the clinical phenotype but there is little phylogenetic context. Regions of intermediate variability have also been suggested for typing. A sequence-based ribo-typing method that uses sequence variation in the 16S-23S intergenic spacer region shows promise and, in conjunction with virulence typing, has been applied for subtyping within the *S.* Enteritidis serotype [52, 53].

Serotype recognition is straight forward and many techniques have been developed and discussed in pervious reviews. The bottom line is that any method which gives you a phylogenetic signal will correctly identify the "good" serotypes, which is the vast majority of serotypes. To try to recreate the Kaufman and White scheme with a molecular scheme seems pointless however, when some serotypes are clearly collections of unrelated strains, unrelated by ancestry and host range. However, for backwards compatibility it is important to be able to recognise the Kaufman and White serotypes whilst naming the strains and assigning them to more appropriate groups. Whole genome sequencing will allow this [54]. Assessing variation in the core genome (genes shared by most individuals being typed) allows a phylogenetic classification and the population will cluster at the serotype level. However, valid concerns around the choice of core genes remain. For MLST seven conserved genes are used but there are 69 *Salmonella* specific genes and so the sample could be expanded but still allow all *Salmonella* to be typed [55]. Leekitcharoenphon [54] used the 69 conserved genes to create a consensus tree; in this tree the *S. enterica* sub-species *arizonae* (an entirely separate sub-species from *enterica*) clustered with the *S. enterica* subspecies *enterica* strains. Their conclusion that the variation in the 69 genes present across *Salmonella* distorts the true differences between the sub-species (they are conserved after all) and so should not be used for typing at this level. Instead, they used an alternative approach, known as pan-genome analysis i.e. the absence of genes, as well as the variation in those present, was used to assess genome similarity. The method worked but some are concerned that the quantitative nature of the comparison may be lost. Where the differences (as between serotypes) are clear then this approach gives a more robust tree (boot strap values for deep branches are typically very strong) and the reduction in computational power looking for presence or absence is very attractive to the reference laboratory. There is, however, a trade-off; the impact on the tree of the presence or absence of a gene is not related to the number of genetic events. A large deletion event removing 20 genes or so would have a much higher impact than a small deletion creating a pseudogene and yet both represent a single event. On the other hand, the relative occurrence of quantifiable single genetic events, whilst excellent for studies of evolution, as it represents the molecular clock and so the time since the last common ancestor for two strains, may not represent the differences that we need to assess public health strategy. The algorithms written to handle genomic data need to be carefully thought through and a welcome development is the suggestion for utilisation of standard operating procedures [56]. The methods need to be used more in real situations in order to assess their discriminatory power and reproducibility, and so determine the utility of the clustering generated.

An alternative approach, kSNP [57] compares similarity between regions k base pairs long in different strains and locates them on a reference genome, not to define them as SNPs but rather to allow the correct k-mer to be identified. This method is not influenced by errors in the reference genome where every isolates seems to have a SNP compared to the erroneous base called in the reference. This method was used to investigate the variation in the *S. enterica* population through time showing that most diversity is recently acquired in a very similar pattern to the *S.* Paratyphi A population. The method can be automated and seems to suggest a way forward for analysing the tens of thousands of genomes being generated by national reference laboratories.

Future Directions for *Salmonella* Investigations

Intestinal infectious disease reference laboratories, from the USA and the UK, are now using whole genomes sequencing of isolates as the method of choice for identifying bacterial pathogens and for sub-typing during outbreak investigation. The methods are quicker, cheaper and more information rich than the current typing methods. The issues of how to analyse the data to give the most reliable identification will inevitably involve the rewriting of the Kaufman-White scheme for serotype designation. The, now dated, method of MLST gave us a glimpse at how this might be achieved through phylogenetically relevant methods but also opened our eyes to the importance of understanding the biology behind the variation as well as the impact of the variation on the biology of the strains. Genetic variation for *Salmonella* is not a continuum of mutations accumulated over time. Lineage-through-time plots show periods of elevated serotype diversification followed by long periods of relatively lower but constant diversification [42, 57]. On a phylogenetic tree this manifests as several early branches which extend and then bificate into the leaves near the tips of the lines. The explanation for this pattern has come from work on *S.* Paratyphi A—the same pattern is seen within this serotype [42]. Variation arises which may allow niche exploitation and the increase in population size provides a very strong positive selection for that mutation such that the change sweeps through the entire population other mutations confer an advantage only under certain circumstance (e.g. resistance to toxins), some have a cost and some are neutral. Many mutations are acquired somewhere in the growing population and most are weeded out through the purifying selection of competition which reduces the frequency of strains with mutations that reduce biological fitness. Recently acquired variation remains extant in the population until removed by purifying selection or allowed to sweep the entire population—so that strains with the new allele replace all other strains. Mutations which have no cost and no benefit accumulate at the rate of the molecular clock and can act as a marker of age. We need to understand these mechanisms of the generation of genetic variation and the impact on biological fitness, then we can exploit fully the massive opportunity afforded by while genome sequencing of every referred isolate.

Antibiotic resistance is yet another phenotype under variable selection pressure. The presence of a resistance phenotype can confer absolute advantage (survival vs. death) in the presence of the antibiotic or have no impact if the antibiotic to which resistance is encoded is not used—long term survival of antibiotic resistance genes depends on "selfish" behaviour and requires an advantage to be conferred, or least no cost, in the absence of the antibiotic. If a biological cost is incurred, not constant for all genes nor mobile elements, and the yet the gene remains it he population then the gene could be an addiction system (cost to the host on loss of the gene) [58]. In order to best manage antibiotic use then susceptibility/resistance needs to be assessed as a continuous phenotype to monitor trends rather than a dichotomous variable based on the prediction of treatment outcome. This will require new techniques to be developed alongside the new sequence based identification and typing methods.

The naming of isolates has to be clinically and biologically relevant in terms of niche occupation as well as in terms of ancestry, all we need is an accurate sequence of the entire genome and an understanding of the biology of the variation seen in the sequence of the population. Perhaps the next step in the development of identification and typing tools is the generation of sequence directly from clinical samples [59]. Nanopore technology promises long reads at an affordable price on a machine which can be used in any laboratory [51]. Currently Pacific Biosytems long read technology allows genomes to be assembled into a single read for a chromosome or a plasmid [60]. If nanopore technology can reach the same quality of data, or Pacific Biosystems launch a benchtop machine, then whole chromosomes could be generated directly from stool and fed into the systems currently under development for sequence analysis. Although currently a research tool advances in nucleic acid extraction and detection protocols and particularly in pathogen DNA enrichment promise great advances in infectious disease control [61]. Coupled with new DNA sequencing technology this would generate data in real time for clinical management of humans and animals, and for outbreak investigation and public health management.

References

1. Fookes M, Schroeder GN, Langridge GC, Blondel CJ, Mammina C, Connor TR, Seth-Smith H, Vernikos GS, Robinson KS, Sanders M, Petty NK, Kingsley RA, Baumler AJ, Nuccio SP, Contreras I, Santiviago CA, Maskell D, Barrow P, Humphrey T, Nastasi A, Roberts M, Frankel G, Parkhill J, Dougan G, Thomson NR. *Salmonella bongori* provides insights into the evolution of the Salmonellae. PLoS Pathog. 2011;7, e1002191.
2. Munch S, Wernery U, Kinne J, Joseph M, Braun P, Pees M, Flieger A, Fruth A, Rabsch W. Comparing the presence of different genes in Salmonella subspecies I–IV and development of a diagnostic multiplex PCR method for identification of Salmonella subspecies. Berl Munch Tierarztl Wochenschr. 2013;126:16–24.
3. Nair S, Wain J, Connell S, de Pinna E, Peters T. *Salmonella enterica* subspecies II infections in England and Wales—the use of multilocus sequence typing to assist serovar identification. J Med Microbiol. 2014;63:831–4.

4. Zhou Z, McCann A, Litrup E, Murphy R, Cormican M, Fanning S, Brown D, Guttman DS, Brisse S, Achtman M. Neutral genomic microevolution of a recently emerged pathogen, *Salmonella enterica* serovar Agona. PLoS Genet. 2013;9, e1003471.
5. Kingsley RA, Baumler AJ. Host adaptation and the emergence of infectious disease: the Salmonella paradigm. Mol Microbiol. 2000;36:1006–14.
6. Holt KE, Thomson NR, Wain J, Langridge GC, Hasan R, Bhutta ZA, Quail MA, Norbertczak H, Walker D, Simmonds M, White B, Bason N, Mungall K, Dougan G, Parkhill J. Pseudogene accumulation in the evolutionary histories of Salmonella enterica serovars Paratyphi A and Typhi. BMC Genomics. 2009;10:36.
7. Scallan E, Hoekstra RM, Angulo FJ, Tauxe RV, Widdowson MA, Roy SL, Jones JL, Griffin PM. Foodborne illness acquired in the United States—major pathogens. Emerg Infect Dis. 2011;17:7–15.
8. EFSA. The community summary report on trends and sources of zoonoses, zoonotic agents, antimicrobial resistance and foodborne outbreaks in the European Union in 2006. 2007; 130. http://www.efsa.europa.eu/en/scdocs/scdoc/130r.htm.
9. Achtman M. Evolution, population structure, and phylogeography of genetically monomorphic bacterial pathogens. Annu Rev Microbiol. 2008;62:53–70.
10. Hopkins KL, Peters TM, de Pinna E, Wain J. Standardisation of multilocus variable-number tandem-repeat analysis (MLVA) for subtyping of Salmonella enterica serovar Enteritidis. Euro Surveill. 2011;16:2–12.
11. Wood JD, Chalmers GA, Fenton RA, Pritchard J, Schoonderwoerd M, Lichtenberger WL. Persistent shedding of Salmonella enteritidis from the udder of a cow. Can Vet J. 1991;32:738–41.
12. Volf J, Stepanova H, Matiasovic J, Kyrova K, Sisak F, Havlickova H, Leva L, Faldyna M, Rychlik I. Salmonella enterica serovar Typhimurium and Enteritidis infection of pigs and cytokine signalling in palatine tonsils. Vet Microbiol. 2012;156:127–35.
13. Bartholomew ML, Heffernan RT, Wright JG, Klos RF, Monson T, Khan S, Trees E, Sabol A, Willems RA, Flynn R, Deasy MP, Jones B, Davis JP. Multistate outbreak of Salmonella enterica serotype enteritidis infection associated with pet guinea pigs. Vector Borne Zoonotic Dis. 2014;14:414–21.
14. Nauerby B, Pedersen K, Dietz HH, Madsen M. Comparison of Danish isolates of Salmonella enterica serovar enteritidis PT9a and PT11 from hedgehogs (*Erinaceus europaeus*) and humans by plasmid profiling and pulsed-field gel electrophoresis. J Clin Microbiol. 2000;38:3631–5.
15. Grzymajlo K, Ugorski M, Kolenda R, Kedzierska A, Kuzminska-Bajor M, Wieliczko A. FimH adhesin from host unrestricted Salmonella Enteritidis binds to different glycoprotein ligands expressed by enterocytes from sheep, pig and cattle than FimH adhesins from host restricted Salmonella Abortus-ovis, Salmonella Choleraesuis and Salmonella Dublin. Vet Microbiol. 2013;166:550–7.
16. van Duijkeren E, Wannet WJ, Houwers DJ, van Pelt W. Serotype and phage type distribution of salmonella strains isolated from humans, cattle, pigs, and chickens in the Netherlands from 1984 to 2001. J Clin Microbiol. 2002;40:3980–5.
17. Heilbronn C, Munnoch S, Butler MT, Merritt TD, Durrheim DN. Timeliness of Salmonella Typhimurium notifications after the introduction of routine MLVA typing in NSW. N S W Public Health Bull. 2014;24:159–63.
18. Brown DJ, Baggesen DL, Hansen HB, Hansen HC, Bisgaard M. The characterization of Danish isolates of *Salmonella enterica* serovar Enteritidis by phage typing and plasmid profiling: 1980–1990. Acta Pathol Microbiol Immunol Scand. 1994;102:208–14.
19. Peters TM, Berghold C, Brown D, Coia J, Dionisi AM, Echeita A, Fisher IS, Gatto AJ, Gill N, Green J, Gerner-Smidt P, Heck M, Lederer I, Lukinmaa S, Luzzi I, Maguire C, Prager R, Usera M, Siitonen A, Threlfall EJ, Torpdahl M, Tschape H, Wannet W, Zwaluw WK. Relationship of pulsed-field profiles with key phage types of *Salmonella enterica* serotype Enteritidis in Europe: results of an international multi-centre study. Epidemiol Infect. 2007;135:1274–81.

20. Boxrud D, Pederson-Gulrud K, Wotton J, Medus C, Lyszkowicz E, Besser J, Bartkus JM. Comparison of multiple-locus variable-number tandem repeat analysis, pulsed-field gel electrophoresis, and phage typing for subtype analysis of *Salmonella enterica* serotype Enteritidis. J Clin Microbiol. 2007;45:536–43.
21. Achtman M, Wain J, Weill FX, Nair S, Zhou Z, Sangal V, Krauland MG, Hale JL, Harbottle H, Uesbeck A, Dougan G, Harrison LH, Brisse S, Group SEMS. Multilocus sequence typing as a replacement for serotyping in *Salmonella enterica*. PLoS Pathog. 2012;8, e1002776.
22. Deng X, Desai PT, den Bakker HC, Mikoleit M, Tolar B, Trees E, Hendriksen RS, Frye JG, Porwollik S, Weimer BC, Wiedmann M, Weinstock GM, Fields PI, McClelland M. Genomic epidemiology of *Salmonella enterica* serotype Enteritidis based on population structure of prevalent lineages. Emerg Infect Dis. 2014;20:1481–9.
23. Winslow CE, Kligler IJ, Rothberg W. Studies on the classification of the Colon-Typhoid Group of Bacteria with special reference to their fermentative reactions. J Bacteriol. 1919;4:429–503.
24. Thomson NR, Clayton DJ, Windhorst D, Vernikos G, Davidson S, Churcher C, Quail MA, Stevens M, Jones MA, Watson M, Barron A, Layton A, Pickard D, Kingsley RA, Bignell A, Clark L, Harris B, Ormond D, Abdellah Z, Brooks K, Cherevach I, Chillingworth T, Woodward J, Norberczak H, Lord A, Arrowsmith C, Jagels K, Moule S, Mungall K, Sanders M, Whitehead S, Chabalgoity JA, Maskell D, Humphrey T, Roberts M, Barrow PA, Dougan G, Parkhill J. Comparative genome analysis of Salmonella Enteritidis PT4 and Salmonella Gallinarum 287/91 provides insights into evolutionary and host adaptation pathways. Genome Res. 2008;18:1624–37.
25. Langridge GC, Fookes M, Connor TR, Feltwell T, Feasey N, Parsons BN, Seth-Smith HM, Barquist L, Stedman A, Humphrey T, Wigley P, Peters SE, Maskell DJ, Corander J, Chabalgoity JA, Barrow P, Parkhill J, Dougan G, Thomson NR. Patterns of genome evolution that have accompanied host adaptation in Salmonella. Proc Natl Acad Sci U S A. 2015;112:863–8.
26. Zheng J, Pettengill J, Strain E, Allard MW, Ahmed R, Zhao S, Brown EW. Genetic diversity and evolution of *Salmonella enterica* serovar Enteritidis strains with different phage types. J Clin Microbiol. 2014;52:1490–500.
27. Bakker HC, Switt AI, Cummings CA, Hoelzer K, Degoricija L, Rodriguez-Rivera LD, Wright EM, Fang R, Davis M, Root T, Schoonmaker-Bopp D, Musser KA, Villamil E, Waechter H, Kornstein L, Furtado MR, Wiedmann M. A whole-genome single nucleotide polymorphism-based approach to trace and identify outbreaks linked to a common *Salmonella enterica* subsp. enterica serovar Montevideo pulsed-field gel electrophoresis type. Appl Environ Microbiol. 2011;77:8648–55.
28. Allard MW, Luo Y, Strain E, Li C, Keys CE, Son I, Stones R, Musser SM, Brown EW. High resolution clustering of *Salmonella enterica* serovar Montevideo strains using a next-generation sequencing approach. BMC Genomics. 2012;13:32.
29. Hopkins KL, de Pinna E, Wain J. Prevalence of Salmonella enterica serovar 4,[5],12:i:—in England and Wales, 2010. Euro Surveill. 2012;17(37), pii: 1–16.
30. Sintchenko V, Wang Q, Howard P, Ha CW, Kardamanidis K, Musto J, Gilbert GL. Improving resolution of public health surveillance for human *Salmonella enterica* serovar Typhimurium infection: 3 years of prospective multiple-locus variable-number tandem-repeat analysis (MLVA). BMC Infect Dis. 2012;12:78.
31. Wain J, Keddy KH, Hendriksen RS, Rubino S. Using next generation sequencing to tackle non-typhoidal Salmonella infections. J Infect Dev Ctries. 2013;7:1–5.
32. EFSA Panel on Biological Hazards (BIOHAZ). Scientific opinion on monitoring and assessment of the public health risk of "Salmonella Typhimurium-like" strains. EFSA J. 2010;8:1826.
33. Octavia S, Wang Q, Tanaka MM, Kaur S, Sintchenko V, Lan R. Delineating community outbreaks of *Salmonella enterica* serovar Typhimurium using whole genome sequencing: insights into genomic variability within an outbreak. J Clin Microbiol. 2015;53(4):1063–71.
34. Okoro CK, Kingsley RA, Connor TR, Harris SR, Parry CM, Al-Mashhadani MN, Kariuki S, Msefula CL, Gordon MA, de Pinna E, Wain J, Heyderman RS, Obaro S, Alonso PL, Mandomando I, MacLennan CA, Tapia MD, Levine MM, Tennant SM, Parkhill J, Dougan

G. Intracontinental spread of human invasive Salmonella Typhimurium pathovariants in sub-Saharan Africa. Nat Genet. 2012;44:1215–21.
35. Phan MD, Wain J. IncHI plasmids, a dynamic link between resistance and pathogenicity. J Infect Dev Ctries. 2008;2:272–8.
36. Peters T, Hopkins KL, Lane C, Nair S, Wain J, de Pinna E. Emergence and characterization of *Salmonella enterica* serovar Typhimurium phage type DT191a. J Clin Microbiol. 2010;48:3375–7.
37. Bone A, Noel H, Le Hello S, Pihier N, Danan C, Raguenaud ME, Salah S, Bellali H, Vaillant V, Weill FX, Jourdan-da Silva N. Nationwide outbreak of *Salmonella enterica* serotype 4,12:i:- infections in France, linked to dried pork sausage, March–May 2010. Euro Surveill. 2010;15:2–4.
38. Mossong J, Marques P, Ragimbeau C, Huberty-Krau P, Losch S, Meyer G, Moris G, Strottner C, Rabsch W, Schneider F. Outbreaks of monophasic *Salmonella enterica* serovar 4,[5],12:i:- in Luxembourg, 2006. Euro Surveill. 2007;12:E11–2.
39. Agasan A, Kornblum J, Williams G, Pratt CC, Fleckenstein P, Wong M, Ramon A. Profile of *Salmonella enterica* subsp. enterica (subspecies I) serotype 4,5,12:i:- strains causing food-borne infections in New York City. J Clin Microbiol. 2002;40:1924–9.
40. Echeita MA, Herrera S, Usera MA. Atypical, fljB-negative *Salmonella enterica* subsp. enterica strain of serovar 4,5,12:i:- appears to be a monophasic variant of serovar Typhimurium. J Clin Microbiol. 2001;39:2981–3.
41. Trupschuch S, Laverde Gomez JA, Ediberidze I, Flieger A, Rabsch W. Characterisation of multidrug-resistant Salmonella Typhimurium 4,[5],12:i:- DT193 strains carrying a novel genomic island adjacent to the thrW tRNA locus. Int J Med Microbiol. 2010;300:279–88.
42. Zhou Z, McCann A, Weill FX, Blin C, Nair S, Wain J, Dougan G, Achtman M. Transient Darwinian selection in *Salmonella enterica* serovar Paratyphi A during 450 years of global spread of enteric fever. Proc Natl Acad Sci U S A. 2014;111:12199–204.
43. Didelot X, Achtman M, Parkhill J, Thomson NR, Falush D. A bimodal pattern of relatedness between the Salmonella Paratyphi A and Typhi genomes: convergence or divergence by homologous recombination? Genome Res. 2007;17:61–8.
44. Kidgell C, Reichard U, Wain J, Linz B, Torpdahl M, Dougan G, Achtman M. Salmonella typhi, the causative agent of typhoid fever, is approximately 50,000 years old. Infect Genet Evol. 2002;2:39–45.
45. Roumagnac P, Weill FX, Dolecek C, Baker S, Brisse S, Chinh NT, Le TA, Acosta CJ, Farrar J, Dougan G, Achtman M. Evolutionary history of Salmonella typhi. Science. 2006;314:1301–4.
46. Wain J, Kidgell C. The emergence of multidrug resistance to antimicrobial agents for the treatment of typhoid fever. Trans R Soc Trop Med Hyg. 2004;98:423–30.
47. Butler T, Rumans L, Arnold K. Response of typhoid fever caused by chloramphenicol-susceptible and chloramphenicol-resistant strains of Salmonella typhi to treatment with trimethoprim-sulfamethoxazole. Rev Infect Dis. 1982;4:551–61.
48. Bhutta ZA, Khan IA, Shadmani M. Failure of short-course ceftriaxone chemotherapy for multidrug-resistant typhoid fever in children: a randomized controlled trial in Pakistan. Antimicrob Agents Chemother. 2000;44:450–2.
49. Wain J, Diep TS, Ho VA, Walsh AM, Nguyen TT, Parry CM, White NJ. Quantitation of bacteria in blood of typhoid fever patients and relationship between counts and clinical features, transmissibility, and antibiotic resistance. J Clin Microbiol. 1998;36:1683–7.
50. Holt KE, Phan MD, Baker S, Duy PT, Nga TV, Nair S, Turner AK, Walsh C, Fanning S, Farrell-Ward S, Dutta S, Kariuki S, Weill FX, Parkhill J, Dougan G, Wain J. Emergence of a globally dominant IncHI1 plasmid type associated with multiple drug resistant typhoid. PLoS Negl Trop Dis. 2011;5, e1245.
51. Ashton PM, Nair S, Dallman T, Rubino S, Rabsch W, Mwaigwisya S, Wain J, O'Grady J. MinION nanopore sequencing identifies the position and structure of a bacterial antibiotic resistance island. Nat Biotechnol. 2014;33(3):296–300.

52. Morales CA, Gast R, Guard-Bouldin J. Linkage of avian and reproductive tract tropism with sequence divergence adjacent to the 5S ribosomal subunit rrfH of Salmonella enterica. FEMS Microbiol Lett. 2006;264:48–58.
53. Morales CA, Musgrove M, Humphrey TJ, Cates C, Gast R, Guard-Bouldin J. Pathotyping of *Salmonella enterica* by analysis of single-nucleotide polymorphisms in cyaA and flanking 23S ribosomal sequences. Environ Microbiol. 2007;9:1047–59.
54. Leekitcharoenphon P, Lukjancenko O, Friis C, Aarestrup FM, Ussery DW. Genomic variation in *Salmonella enterica* core genes for epidemiological typing. BMC Genomics. 2012;13:88.
55. Lukjancenko O, Ussery D. Design of an Enterobacteriaceae Pan-Genome Microarray Chip. In: Chan J, Ong Y-S, Cho S-B, editors. Computational systems-biology and bioinformatics. Berlin: Springer; 2010. p. 165–79.
56. Snipen L, Ussery DW. Standard operating procedure for computing pangenome trees. Stand Genomic Sci. 2010;2:135–41.
57. Timme RE, Pettengill JB, Allard MW, Strain E, Barrangou R, Wehnes C, Van Kessel JS, Karns JS, Musser SM, Brown EW. Phylogenetic diversity of the enteric pathogen *Salmonella enterica* subsp. enterica inferred from genome-wide reference-free SNP characters. Genome Biol Evol. 2013;5:2109–23.
58. Deane SM, Rawlings DE. Plasmid evolution and interaction between the plasmid addiction stability systems of two related broad-host-range IncQ-like plasmids. J Bacteriol. 2004;186:2123–33.
59. Wain J, Mavrogiorgou E. Next-generation sequencing in clinical microbiology. Expert Rev Mol Diagn. 2013;13:225–7.
60. Liao YC, Lin SH, Lin HH. Completing bacterial genome assemblies: strategy and performance comparisons. Sci Rep. 2015;5:8747.
61. Rubino S, Wain J, Gaind R, Paglietti B. A novel broadly applicable PCR-RFLP method for rapid identification and subtyping of H58 Salmonella Typhi. J Microbiol Methods. 2016;127:219–23.

Chapter 7
Verocytotoxin-Producing *Escherichia coli* in the Genomic Era: From Virulotyping to Pathogenomics

Valeria Michelacci, Rosangela Tozzoli, Alfredo Caprioli, and Stefano Morabito

Verocytotoxin Producing *E. coli*: From the Origin to the "Next" Era

Escherichia coli is an ubiquitous bacterial species representing an important component of the microbiota in both human and animal hosts. At the same time, it is one of the most diffuse bacterial species in many environmental niches, including surface water and soil. Horizontal gene transfer (HGT) played a key role in determining the success of *E. coli* as ubiquitous bacterial species. Mobile genetic elements (MGE), including plasmids, bacteriophages and pathogenicity islands, accumulated into the *E. coli* genome and favoured the selection of the most suitable individuals to colonize new ecological niches.

In the human host, *E. coli* strains are commensals and exert a beneficial effect. However, some individuals evolved the capability to harm and cause disease, following the acquisition of virulence determinants through the HGT.

Verocytotoxin-producing *E. coli* are iconic of this evolutionary pathway. In late 1970s, Konowalchuck and colleagues discovered that certain *Escherichia coli* strains produced a potent cytotoxin inducing a cytopathic effect on Vero cells monolayers [1, 2]. The toxin was named "Vero cell toxin" and the bacteria "verocytotoxin-producing *E. coli*" (VTEC). In 1983 an *E. coli* strain belonging to a rare serotype, O157:H7, was recognized as the causative agent of haemorrhagic colitis and haemolytic uremic syndrome during the investigations on two outbreak episodes [3, 4]. Since then, VTEC belonging to the same serotype have become increasingly common as food-borne pathogens and nowadays O157:H7 is one of the most common VTEC serotypes causing severe disease in humans, followed by VTEC from a dozen

V. Michelacci (✉) • R. Tozzoli • A. Caprioli • S. Morabito
European Union Reference Laboratory for Escherichia coli,
Department of Veterinary Public Health and Food Safety, Istituto Superiore di Sanità,
viale Regina Elena 299, 00161 Rome, Italy
e-mail: valeria.michelacci@iss.it

X. Deng et al. (eds.), *Applied Genomics of Foodborne Pathogens*, Food Microbiology and Food Safety, DOI 10.1007/978-3-319-43751-4_7

of additional serogroups, including O26, O111, O103, O145, O121 and O45 among others [5–8], altogether termed as enterohaemorragic *E. coli* (EHEC) [9, 10].

VTEC O157 and possibly other VTEC serogroups are zoonotic pathogens. Ruminants have been recognized as their natural reservoir and excrete the pathogen into the environment with their faeces. Human infections occur via ingestion of raw or undercooked contaminated food and via contact with colonized animals and contaminated environment.

Clinical manifestations of VTEC infection range from asymptomatic carriage to uncomplicated diarrhoea, haemorrhagic colitis (HC) and the life-threatening haemolytic uremic syndrome (HUS) [6, 9, 11]. This wide spectrum of symptoms, in addition to difficulties in the diagnostic procedures used to identify these pathogens [12], causes these infections to be overlooked, especially the less severe forms, making the burden of VTEC disease largely underestimated. Culture-based detection of pathogens from either clinical specimens or food samples usually relies on the availability of selective and/or differential media, but VTEC detection can't benefit from such an approach. As a matter of fact, VTEC are mostly indistinguishable from commensal *E. coli* by culturing with the exception of *E. coli* O157:H7, which generally possess peculiar phenotypic properties (inability to ferment sorbitol, absence of beta-glucuronidase activity, resistance to cefixime and potassium tellurite).

Beside the diagnosis of infections and food testing, the inability to distinguish between pathogenic and harmless *E. coli* strains affects the surveillance and the monitoring of this pathogen. For a VTEC strain to be characterised, the most meaningful approach resides in the identification of virulence genes, or the so-called "virulome" [13, 14]. Such genes include the determinants encoding the Verocytotoxins (VT) and several other virulence factors involved in the efficient colonization of VTEC in host gut [13, 15–18]. The definition of the VTEC virulome is still an ongoing process. It started soon after their discovery with the identification of the bacteriophages conveying the VT-coding genes [18] and continued with the identification of many accessory pathogenicity islands, such as the LEE locus [16] and the O-islands (OI) 122 [15, 17] and 57 [13]. Nevertheless, the VTEC virulome seems to be even more complex both in its size and pathogenetic potential, as it is witnessed by the genomic information derived from the determination of the first whole genome sequences of VTEC O157 [19, 20] and the advent of the "Next Generation Sequencing" era.

VTEC Pathogenomics and the Problem of Its Classification

VTEC is a highly heterogeneous group of pathogenic *E. coli*, comprising strains with different genetic and phenotypic features, and characterised by diverse combinations of virulence determinants, which may be related to the observed wide range of symptoms associated with VTEC disease.

VT-phages are the main virulence-associated MGE of the VTEC virulome and have a high degree of variation in their genomes [21]. Two main types of VTs have been recognized, VT1 and VT2 [22], which are antigenically distinct [22, 23]. The identification of VT-genes (*vtx*) allowed the development of molecular tools for the detection of VTEC [18, 24–26], which involved the use of DNA probes at the beginning but later transitioned to PCR amplification of specific parts of the VT-coding genes [27, 28]. Nowadays, this approach has been upgraded with tools able to identify the numerous subtypes and variants of each main VT type, including three subtypes of VT1 (VT1a, VT1c and VT1d) and seven subtypes of VT2 (VT2a, VT2b, VT2c, VT2d, VT2e, VT2f and VT2g) [29, 30], constituting a means to better characterise VTEC strains. As a matter of fact, a VTEC strain can produce either VT1 or VT2 alone or both in different combinations of type/subtype, even though some of these can be more frequent in certain VTEC subpopulations. In particular, several studies have indicated that the presence of some *vtx2* genes subtypes, namely *vtx2a* and *vtx2c*, is strongly associated with VTEC strains causing serious illness [29, 31].

Although the VTs represent the major virulence trait of VTEC, their sole production seems to be not sufficient to cause illness, at least the most severe forms. Indeed, for the disease to become apparent, the effective colonization of the host gut must be ensured. Most of the VTEC associated with HC or HUS, also termed EHEC, produce a typical hystopathological lesion to enterocytes termed "attaching and effacing" (A/E). A/E is characterized by the effacement of the microvilli brush border and the presence of microfilamentous structures, known as pedestals, which protrude from the cell surface and intimately accommodate the adhered bacteria [16]. The ability to cause the A/E lesion is governed by several genes present on a 35 Kbp pathogenicity island called the Locus of Enterocyte Effacement (LEE), which was first described in Enteropathogenic *E. coli* (EPEC) [32], another group of diarrheagenic *E. coli*, and later shown to be present in other bacterial pathogens, including EHEC and bacteria belonging to other enterobacteriaceae species, such as *Citrobacter freundii*, *Hafnia alvei* and *Escherichia albertii* [33–36].

The LEE locus conveys the genes encoding a type III secretion system (TTSS), a complex molecular structure allowing the injection of bacterial effectors directly into the host cell, and the genes *eaeA* and *tir* encoding an outer membrane protein called intimin, which mediates the intimate attachment of the bacterium to the enterocyte surface, and its translocated receptor, respectively [37]. As with the VT-genes, the *eaeA* gene presents a considerable diversity in the nucleotide sequence [33, 38, 39], resulting in several distinct intimin types classified with a nomenclature system based on the Greek alphabet [40, 41]. Some intimin types are generally associated with certain VTEC groups, such as intimin γ produced by serogroups O157, O111, and O145, and intimin ε possessed by VTEC serogroups O103 and O121 [41]. It has been hypothesised that the different intimin types confer to the *E. coli* strains different tissue tropism. In fact EPEC strains, which produce β intimin, have been shown to colonise almost all regions of the small bowel, while the colonization of γ intimin-positive EHEC seems to be restricted to the follicle-associated epithelium of the Peyer's patches [42].

It has been suggested that LEE-positive VTEC such as the EHEC serotypes O157:H7, O26:H11, O103:H2, O111:NM, O121:H19, and O145:NM are more commonly associated with HUS and outbreaks than LEE-negative VTEC serotypes [10,

15, 43]. This assumption seems not to be absolute though. As a matter of fact, some LEE-positive VTEC identified in animal reservoirs belong to serogroups that have never been detected in human cases of disease [44, 45], while VTEC lacking this pathogenicity island have been isolated from cases of HC and HUS [46–49]. These observations suggest that, in addition to the LEE locus, other genetic elements may also constitute the VTEC virulome.

Unravelling the complete mechanism of VTEC pathogenesis and the complex structure of VTEC genomes is therefore pivotal to a comprehensive definition of the strains that may be considered as pathogenic to human beings.

The Discovery of the Mosaic Nature of VTEC Chromosome and the Attempts to Define Pathogenic VTEC in the Pre-NGS Era

In the early 2000s, Hayashi et al. [19] published the whole genome sequence (WGS) of the EHEC O157:H7 RIMD 0509952 strain (also termed O157 Sakai), which had caused a huge outbreak of infections in 1996 in Japan affecting more than 6000 schoolchildren [50]. In the same period, the genome of the O157:H7 strain EDL933, isolated from Michigan ground beef and linked to a multi-state outbreak of HC in the United States in 1982 [4], was sequenced and annotated [20]. These studies revealed that horizontal gene transfer played a far more extensive role in VTEC O157 evolution than expected [19, 20, 51] and opened the way to the age of VTEC genomics. The analysis of the two VTEC genomes in comparison with that of the *E. coli* K12 reference strain MG1655 [52] revealed for the former a genome of more than 5 Mbp and highlighted the existence of a 4.1 Mbp backbone shared between MG1655 and EDL933. The VTEC O157 genome contained 177 unique "O-islands" (OI) while the K12 MG1655 strain had 234 "K-islands" [20]. A similar picture was depicted for the RIMD 0509952 strain, whose chromosome is 5.4 Mbp in length, with similar presence of backbone and strain-specific segments [19]. As a whole, about 20 % of the VTEC O157 genome was specific to individual strains. This large amount of accessory DNA comprises both the main known virulence-associated genetic elements, such as the LEE locus and the VT-converting phages, and many other MGEs that could encode additional virulence factors or other properties involved in the pathogenetic mechanism or the survival of the bacterium in the environment or the food chain. As a matter of fact, only 40 % of the open reading frames (ORFs) identified in the OI of the VTEC O157 EDL933 strain has been assigned with a putative function [20]. In the EDL933 strain, apart from the LEE locus, some other OIs larger than 15 Kbp have been regarded as pathogenicity islands (PAIs) since they encode putative virulence factors, have a GC content lower than the average of the *E. coli* chromosome, and are inserted in or close to tRNA loci [53]. In particular, the attention of the investigators was drawn by a 22-Kbp genomic island, designated as OI-122 in strain EDL933 [20]. This island contained the 5' region of the *efa1/lifA* gene, whose product has been described to reduce the immune response of the host upon EPEC infection and to be involved in the ability to adhere to cultured epithelial cells [54, 55]. After the identification of the *efa1/lifA* gene, two independent studies investigated the role of OI-122 in VTEC evolution [15, 17] and came to the conclusion that this OI has co-evolved with the LEE locus in VTEC and EPEC strains [17] and that it is part of the most pathogenic VTEC virulome [15].

The growing knowledge of the VTEC virulome led to the concept of "seropathotype" (SPT) [15], which became a means to define different VTEC types. It consisted in grouping VTEC serotypes based on their reported frequencies in human illness (in qualitative terms such as "high," "moderate," or "rare"), their known associations with outbreaks and with severe disease, such as HUS and HC, and on the presence of the LEE locus and the OI-122 [15]. The derived classification scheme divided VTEC into five SPT (A–E), in a decreasing rank of pathogenicity (Table 7.1, modified from EFSA [12]).

SPT-A comprises serotypes O157:H7 and O157:NM, which are the most common causes of outbreaks and HUS, while SPT-B includes the remaining VTEC of the EHEC group such as serotypes O26:H11, O103:H2, O111:NM, O121:H19 and O145:NM. SPT-C includes LEE-negative VTEC, such as those of serotypes O91:H21 and O113:H21, which are sporadically isolated from HUS. SPT-D contains many VTEC serotypes isolated from diarrhoea but not associated with HUS or HC, while the SPT-E group contains animal isolates that have never been described in human disease [15, 56].

The SPT-based classification of VTEC is helpful to assign most of the VTEC strains isolated from human disease to a category, but it lacks the capacity of proactively assigning a risk rank to VTEC isolated from the vehicles of infections. This is largely due to the many gaps in our knowledge of the features that make a VTEC strain pathogenic and is the subject of the large on-going debate in both scientific

Table 7.1 Classification of VTEC serotypes into seropathotypes

SPT	Incidence	Outbreaks	Severe disease	Virulence markers			Serotypes
				vtx	*eae*	*OI-122*	
A	High	Common	Yes	*vtx*2 but may also carry *vtx*1	+	+	O157:H7, O157:NM
B	Moderate	Uncommon	Yes	*vtx*1 and/or *vtx*2	+	+	O26:H11, O103:H2, O111:NM, O121:H19, O145:NM
C	Low	Rare	Yes	*vtx*1 and/or *vtx*2	+/–	+/–	O91:H21, O104:H21, O113:H21, O5:NM, O121:NM, O165:H25
D	Low	Rare	No	*vtx*1 and/or *vtx*2	+/–	–	Multiple
E	Non-human only	NA	NA	*vtx*1 and/or *vtx*2	+/–	–	Multiple

and regulatory contexts that aims at deploying measures to mitigate the risk of VTEC infection for human beings. As an example, the European Food Safety Authority (EFSA) considered the SPT classification as the basis for defining pathogenic VTEC and identifying the VTEC populations to be monitored in food vehicles to protect consumers' health [12]. The discussion group involved in the production of the opinion analysed the data on human infections collected by the European Centre for Disease prevention and Control (ECDC) in the time period 2007–2010 and concluded that, based on the current knowledge, it was not possible to fully define a human pathogenic VTEC and that probably a dynamic molecular approach taking into consideration the evolving information on the VTEC virulome would better fit the scope of defining the pathogenic VTEC in the future.

Identification of New and Emerging VTEC Through WGS

DNA sequencing technologies have leapfrogged in the recent years towards the production of affordable benchtop instruments to produce DNA sequences from entire genomes, de facto opening the "Next Generation Sequencing" (NGS) era. The genomes of multiple bacterial strains can be simultaneously deciphered and covered at a good depth by operating the current NGS platforms in single runs, producing raw sequencing data in less than 24 h. Even though the NGS technology is continuously improving, some aspects still need to be addressed before the approach can be used as a routine tool for surveillance of pathogenic microorganisms. The length of the sequencing reads remains a crucial factor for minimizing the number of gaps in draft genomes assembled from the raw sequencing reads. Additionally, the presence of DNA regions exchanged among bacteria via HGT hinders the assembly and interpretation of sequencing data, due to the existence, in these MGE, of similar stretches of sequences repeated in multiple positions throughout the genome. As a result, the assembled contigs (i.e., DNA fragments assembled from sequencing reads) are often interrupted at the MGE [57]. This is particularly true when analysing the genomes of pathogenic *Escherichia coli* strains, e.g. VTEC, which are characterized by a high degree of genomic plasticity and contain a large fraction of MGE. In fact, considering the whole population of pathogenic *E. coli*, it has been estimated that approximately 26% of the complete *E. coli* pangenome is made up by "volatile" genes conferring pathotype/strain specificity [58, 59].

The study of the VTEC O104:H4 strain that caused a severe food-borne outbreak of haemorrhagic colitis and haemolytic uremic syndrome in Germany and France in 2011 marked the first application of the NGS technology in the investigation of a multinational outbreak of *E. coli*. The episode involved a total of 4033 cases, comprising 901 HUS and 50 deaths [60, 61]. The outbreak strain was an Enteroaggregative *E. coli* (EAEC) that was able to both display the typical stacked brick pattern of adhesion on cultured cells monolayers and produce Verocytotoxin (VT) [62]. Such unfamiliar combination of virulence features, caused by the acquisition of a VT-converting phage by an EAEC, accounted for the high virulence of the strain as shown by the high number of HUS cases in not immune-compromised adults [63]. Tracing-back investigations have allowed attributing the source of the outbreak to sprouts produced with fenugreek seeds imported from Egypt [61, 64]. The combination of the high pathogenicity of the strain with its accidental dissemination through the food chain resulted in the largest outbreak due to VT-producing *E. coli* ever registered in Europe.

During this outbreak, the value of NGS technology in understanding the virulence, origins and epidemiology of this novel pathogen was readily demonstrated [65–68]. Several research groups sequenced the genome of VTEC O104:H4 strains isolated from patients involved in the outbreak in a few days after the first alert. Of particular interest, a crowdsourcing was set up within days on a public website for sharing the results from the genomic analyses, involving the participation of many European groups [69]. These analyses revealed a strong similarity of the genomic backbone of the outbreak strain with that of an historical O104:H4 EAEC isolate, the strain 55989, with major differences residing in the plasmid content and the exclusive presence of a VT2-coding phage in the outbreak strain [66–68]. In the latter, the VT-coding genes were carried by a lambdoid phage integrated in the *wrbA* locus, a genomic hotspot for phage insertions, and produced the VT2a subtype toxin [66], one of the subtypes most frequently found in VTEC strains isolated from severe cases of disease [31]. The genes encoding the enteroaggregative adhesion phenotype, such as those producing the AAF/I fimbriae, were harboured on a 83 Kbp plasmid similar to that found in classical EAEC strains, suggesting the origin of the outbreak strain from an event of VT-phage acquisition, probably from a bovine source [70], by a typical EAEC. Interestingly, a second large plasmid of 90 Kbp was identified, harbouring the genes conferring the ESBL resistance phenotype [66–68]. The latter trait is becoming a matter of increasing concern in EAEC strains [71] and at the same time an uncommon feature in VTEC strains [72].

During the investigation on the German outbreak of 2011, the genome of another historical *E. coli* O104 strain, HUSEC041, isolated from a sporadic HUS case in Germany in 2001 [66] was also completely sequenced and compared with the 2011 outbreak strain by Multi Locus Sequence Typing using all the identified genes of the core genome (cgMLST). This investigation suggested that the two isolates might belong to a highly pathogenic O104:H4 VT-producing EAEC clone not distantly related to the 55989 reference EAEC O104:H4 strain [66].

The prompt release of the whole genome sequence of the strain responsible for the 2011 German outbreak boosted the development of diagnostic tools. As an example, a novel diagnostic application was invented using the draft whole genome assemblies made available through the *E. coli* O104:H4 Genome Analysis Crowd-Sourcing Consortium website [73]. The PCR-based application involved iterative primers design and alignment predictions. Each primer pair was selected not only by virtue of its ability to recognize sequences on the target genome, but also with respect to its inability to map on the genomes of other 69 *E. coli* strains used as a negative control panel. This application provides a good example of the possibilities offered by the NGS technologies. The proposed opportunity to design tools for detecting specific strains without the need of targeting a predefined genomic feature is promising. It could avoid the need to perform computationally intense and time consuming genome-to-genome alignments while allowing the *in silico* development of diagnostics without a thorough knowledge of the target microorganism [73].

The release of the whole genome sequence of the O104:H4 strain from the 2011 outbreak not only guided the development of diagnostic tools, but also set the basis for comparative studies aiming at investigating the evolutionary mechanisms behind its emergence. The genomic studies converged on the analysis of the sequences of the few other strains available in the public repositories that shared the peculiar characteristics of the O104:H4 outbreak strain.

The first isolation of an EAEC strain producing VT2 dates back to 1992, when a VTEC O111:H10 strain (formerly typed as O111:H2) was reported as being the causative agent of an outbreak of haemolytic uremic syndrome in France [74]. Moreover, after the German outbreak of 2011, two other strains showing the same combination of virulence genes were isolated from as many human cases of disease and subjected to whole genome sequencing. The first isolate, a VT-producing EAEC O111:H21, caused a sporadic HUS case in Northern Ireland in 2012 [75]. The second one, of serotype O127:H4, was isolated from a small HUS outbreak in Italy in 2013 [76]. Recently, it has been shown that the excision of the VT-encoding bacteriophage harboured by the VT-producing EAEC O111:H10 can be induced and that the phage itself can be used for efficaciously infecting *E. coli* strains belonging to different diarrheagenic *E. coli* [76]. The genome of this phage was completely sequenced [77] and showed 99 % of sequence identity with the bacteriophage harbouring the *vtx2* genes in the VT-producing EAEC O104:H4 strain responsible for the German outbreak occurred in 2011 (Acc. No. NC_018846) [78]. A further comparison was carried out between these two phage sequences and the WGS of the other known VT-producing EAEC in order to assess if the VT-phages present in these peculiar VTEC strains shared common features and could thus provide hints on their origin and evolution. Surprisingly, all the VT-phage sequences were almost identical, displaying more than 99 % sequence similarity, with the exception of the VT2-coding bacteriophage harboured by the O111:H21 strain isolated in Northern Ireland [76, 77]. The latter appeared to be completely different from the phages in all the other strains, with the exception of a short DNA stretch of 8 Kbp comprising the *vtx2* genes, providing evidence that at least two different populations of VT-converting phages have been able to stably infect EAEC strains [76, 77].

The finding of a nearly identical VT-phage in EAEC strains belonging to three different serotypes, namely O111:H10, O104:H4 and O127:H4 was puzzling. In fact, since the strains were isolated during a more than 20 year time span and the phages are usually very variable [21], the observation suggested that such VT-phages could be kept under a strong selective pressure impeding the accumulation of sequence variations.

The availability of WGS of VT2-phages from VT-producing EAEC strains allowed their genomic comparison with those from the VT-phages of typical VTEC strains. Such a strategy showed the presence of a short sequence fragment uniquely associated with the phages present in EAEC strains and encoding a tail fiber. As it has been reported that the interactions between phage tail fibers and bacterial host proteins contribute to the success of the infection [79], it is conceivable, although not confirmed yet, that differences in such phage proteins may contribute to define VT-phages tropism for *E. coli* recipients [77].

The studies on VT-phage genomes showed how an NGS-based approach could be funnelled towards the investigation of specific aspects of pathogenic *E. coli* evolution.

As a matter of fact, the use of NGS technology enabled fast characterization of the VT-converting phages harboured by EAEC strains and allowed making hypotheses on the mechanisms underlying the evolution of the VT-producing EAEC.

The results of the studies on EAEC and their VT-phages also ignited the debate on whether VTEC do represent a distinct class of *E. coli* able to produce VTs or rather they have to be considered as variants of other pathogenic *E. coli* groups with augmented pathogenicity, generated by the acquisition of a VT-phage.

NGS: A Promising Subtyping Approach to the Real Time Surveillance and Monitoring of VTEC Infections

The application of NGS to bacterial typing, taking advantage of its impressive discriminatory power, is as attractive as its ability to allow deciphering of the virulence of a bacterial strain.

Molecular typing techniques have been developed and widely used in the last 20 years to characterise bacterial isolates and provided essential tools for surveillance and monitoring of pathogenic bacteria and early detection of outbreaks. This approach has been recently acknowledged and enforced at the EU level for the control and monitoring of food-borne infections caused by pathogens such as *Salmonella*, *Listeria* and VTEC in the perspective of outbreak preparedness [80]. In 2013, the European Food Safety Authority received the mandate from the European Commission to provide technical support for the collection of molecular typing data of food/animal isolates of *Salmonella*, *Listeria monocytogenes* and VTEC. Previously, the European Centre for Disease prevention and Control (ECDC) had been appointed to collect similar information from clinical isolates. The molecular typing information hosted in the two databases will be the basis for a joint investigation of clusters of bacterial profiles involving those from human cases of disease and from non-human sources. The molecular typing data collection program is currently accepting bacterial molecular profiles obtained through standardized typing methods, such as Pulsed Field Gel Electrophoresis (PFGE) and, limited to *Salmonella* isolates, Multi Locus Variable number tandem repeats Analysis (MLVA). Although based on the use of "first generation" typing techniques, those being already standardized and largely in use in the EU, the molecular typing data collection program does not reject the possibility of upgrading to the more modern and sophisticated typing approaches based on WGS in the future [81]. Nevertheless, for their adoption in routine surveillance and monitoring, both the NGS technologies and data analysis should be streamlined and used by a wider spectrum of laboratories.

To respond to the strategic needs related with the introduction of NGS into public health microbiology, ECDC has already appointed an Expert Group on the "Introduction of Next Generation typing methods to surveillance of Food- and Waterborne Disease (FWD)", with the aim of depicting an NGS-based FWD Surveillance Strategic Framework for guiding future implementation of new molecular/genomic typing technologies for surveillance and outbreak investigation of FWD at national and European level. The activity of this group has led to the publication in October 2015 of the "Expert opinion on the introduction of next-generation typing methods for food-and waterborne diseases in the EU and EE" (available at the ECDC webpage: http://ecdc.europa.eu/en/publications/Publications/food-and-waterborne-diseases-next-generation-typing-methods.pdf).

With entire genomes sequenced, the analysis of the variability is extended to single nucleotidic polymorphisms (SNPs) occurring at any position on the microbial genome. This approach increases the discriminatory power of the typing techniques, but may also complicate the interpretation of the results. In order to reduce the resulting complexity of the phylogenetic analyses, whole genome MLST (wgMLST) schemes are currently being developed for several bacterial pathogens, aiming at identifying only the differences (SNPs, insertions and deletions) present in the coding regions [82–84]. For typing some bacterial species the use of a "core genome MLST" (cgMLST) has also been proposed, interrogating the alleles of all the core genes identified upon comparison with a comprehensive species-specific database of core genes only [85–88].

The challenge of using schemes based on the whole genome SNPs (wg-SNPs) for typing VTEC is linked to the need to come to a consensus on the number of SNPs defining a cluster given the high variability characterising this bacterial species [89, 90]. For typing *E. coli* strains, including VTEC, the possibility to use simplified allelic variation patterns, such as those derived from wgMLST or cgMLST, may be more appropriate. It has to be considered, however, that it could be challenging defining either the borders of the different *E. coli* pathotypes' genomes, being many genes often shared among different groups [57, 91], or identifying the core genome of a bacterial species whose pan-genome includes today about 20,000 genes [59].

In 2014 the first attempt at comparing the performance of conventional methods and whole genome sequencing for characterizing and typing VTEC strains was published [89]. The authors demonstrated the suitability of the NGS-based approach in the identification of the virulence genes content and for MLST typing and [92], at the same time observed a cluster of VTEC O157 genomic profiles possibly identifying an outbreak of infections by applying the wg-SNPs typing approach to characterise all the VTEC strains isolated in a 7 weeks sampling period. This observation strengthened the usefulness of a "near real-time" surveillance of VTEC infections through the NGS-based subtyping of the infecting strains with respect to the use of other methods such as serotyping, dot blot-based virulotyping and conventional PFGE that would have led to a similar result but in a much longer period of time [89].

Recently, other studies have explored the application of the SNPs analysis to the investigation of outbreaks of VTEC O157 infections. One of these studies was successful in identifying two different clones of isolates that had caused two distinct outbreaks of infections that had previously been considered as a single episode being the cases reported in the UK in the same period (September 2013) and caused by VTEC O157 belonging to the same phage type (PT2) [93]. The same SNP-based typing approach was also used to retrospectively analyse collections of VTEC O157 strains held at the Public Health England [94, 95]. Such piece of research led to the conclusion that using the SNPs analysis would allow identifying more clusters of isolates, compared to traditional typing techniques [94]. In one of these studies, the authors also attempted to establish a threshold of five SNPs to identify epidemiologically related isolates [94].

An additional study describing an extensive analysis of VTEC O157 circulating in the Great Britain and introducing an interesting algorithm for computing the genomic differences needs to be mentioned. In this piece of research, SNPs were used to describe the population structure of VTEC O157 based on clonal groups. In detail, a hierarchical single linkage clustering was performed on the pairwise SNP difference between the strains at various distance thresholds (Δ250, Δ100, Δ50, Δ25, Δ10, Δ5, Δ0), resulting in unique keys assigned to the SNPs profiles. The key has been termed SNP address and represent the first proposal for a SNP-based nomenclature [96].

Although all the mentioned studies have provided evidences that the SNPs analysis may be useful for outbreak investigation and surveillance studies, it is important to note that most of them were successful in detecting clusters of VTEC O157 strains, which are known to have emerged more recently than VTEC belonging to other serogroups and therefore being much more homogeneous [97], intrinsically reducing the complexity of the typing results.

Only one study based on SNP-typing has been published so far with the purpose of profiling epidemiologically related non-O157 VTEC strains [98]. The isolates investigated were VTEC O26 isolated from two outbreaks with strong epidemiological and microbiological evidence. The authors observed that the isolates from both the episodes showed, within each group, a maximum of three SNPs difference when the VTEC O26 strain 11368 was used as reference for the SNP identification [98]. This finding is interesting but it should be extended to a wider population of strains before considering it as a threshold for the identification of clusters of cases of VTEC O26.

All the aforementioned studies demonstrated the suitability of SNP-based typing for the identification of VTEC O157 and O26 clusters in limited settings but may not prove efficacious in surveillance or monitoring systems where the observations are extended to wider VTEC populations over the time. As a matter of fact, all the proposed approaches were reference-based and only used the portions of the genome in common between a fully annotated reference genome and the isolates assayed for the SNPs identification. This strategy has some limitations. In fact, the genomic plasticity of this bacterial species is expected to cause the field isolates to quickly diverge from the reference used, which would reduce the portion of genome subjected to SNPs identification in turn reducing the discriminatory power of the whole typing system.

More extensive studies should be carried out to assess the suitability of reference-based SNPs analysis for the purpose of replacing the current methodologies used for VTEC typing. Reference-free analytical pipelines should also be developed and evaluated for their ability to produce meaningful typing results in order to increase the possibility for the SNP-typing to convert from a promising approach to a factual tool for the monitoring and the surveillance of VTEC infections.

References

1. Konowalchuk J, Dickie N, Stavric S, Speirs JI. Properties of an *Escherichia coli* cytotoxin. Infect Immun. 1978;20(2):575–7.
2. Konowalchuk J, Speirs JI, Stavric S. Vero response to a cytotoxin of *Escherichia coli*. Infect Immun. 1977;18(3):775–9.
3. Karmali MA, Steele BT, Petric M, Lim C. Sporadic cases of haemolytic-uraemic syndrome associated with faecal cytotoxin and cytotoxin-producing *Escherichia coli* in stools. Lancet. 1983;1(8325):619–20.
4. Riley LW, Remis RS, Helgerson SD, McGee HB, Wells JG, Davis BR, Hebert RJ, Olcott ES, Johnson LM, Hargrett NT, Blake PA, Cohen ML. Hemorrhagic colitis associated with a rare *Escherichia coli* serotype. N Engl J Med. 1983;308(12):681–5. doi:10.1056/NEJM198303243081203.
5. Gould LH, Mody RK, Ong KL, Clogher P, Cronquist AB, Garman KN, Lathrop S, Medus C, Spina NL, Webb TH, White PL, Wymore K, Gierke RE, Mahon BE, Griffin PM, Emerging Infections Program Foodnet Working Group. Increased recognition of non-O157 Shiga toxin-producing *Escherichia coli* infections in the United States during 2000–2010: epidemiologic features and comparison with *E. coli* O157 infections. Foodborne Pathog Dis. 2013;10(5):453–60. doi:10.1089/fpd.2012.1401.
6. Karmali MA. Infection by verocytotoxin-producing *Escherichia coli*. Clin Microbiol Rev. 1989;2(1):15–38.
7. Mathusa EC, Chen Y, Enache E, Hontz L. Non-O157 Shiga toxin-producing *Escherichia coli* in foods. J Food Prot. 2010;73(9):1721–36.
8. Scallan E, Hoekstra RM, Angulo FJ, Tauxe RV, Widdowson MA, Roy SL, Jones JL, Griffin PM. Foodborne illness acquired in the United States—major pathogens. Emerg Infect Dis. 2011;17(1):7–15. doi:10.3201/eid1701.091101p1.
9. Griffin PM, Tauxe RV. The epidemiology of infections caused by *Escherichia coli* O157:H7, other enterohemorrhagic *E. coli*, and the associated hemolytic uremic syndrome. Epidemiol Rev. 1991;13:60–98.
10. Nataro JP, Kaper JB. Diarrheagenic *Escherichia coli*. Clin Microbiol Rev. 1998;11(1):142–201.
11. Karmali MA, Gannon V, Sargeant JM. Verocytotoxin-producing *Escherichia coli* (VTEC). Vet Microbiol. 2010;140(3–4):360–70. doi:10.1016/j.vetmic.2009.04.011.
12. EFSA. Scientific opinion on VTEC-seropathotype and scientific criteria regarding pathogenicity assessment. EFSA J. 2013;11. doi:10.2903/j.efsa.2013.3138.
13. Imamovic L, Tozzoli R, Michelacci V, Minelli F, Marziano ML, Caprioli A, Morabito S. OI-57, a genomic island of *Escherichia coli* O157, is present in other seropathotypes of Shiga toxin-producing *E. coli* associated with severe human disease. Infect Immun. 2010;78(11):4697–704. doi:10.1128/IAI.00512-10.
14. Kohler S, Foulongne V, Ouahrani-Bettache S, Bourg G, Teyssier J, Ramuz M, Liautard JP. The analysis of the intramacrophagic virulome of Brucella suis deciphers the environment encountered by the pathogen inside the macrophage host cell. Proc Natl Acad Sci U S A. 2002;99(24):15711–6. doi:10.1073/pnas.232454299.
15. Karmali MA, Mascarenhas M, Shen S, Ziebell K, Johnson S, Reid-Smith R, Isaac-Renton J, Clark C, Rahn K, Kaper JB. Association of genomic O island 122 of *Escherichia coli* EDL 933 with verocytotoxin-producing *Escherichia coli* seropathotypes that are linked to epidemic and/or serious disease. J Clin Microbiol. 2003;41(11):4930–40.

16. McDaniel TK, Kaper JB. A cloned pathogenicity island from enteropathogenic *Escherichia coli* confers the attaching and effacing phenotype on *E. coli* K-12. Mol Microbiol. 1997;23(2):399–407.
17. Morabito S, Tozzoli R, Oswald E, Caprioli A. A mosaic pathogenicity island made up of the locus of enterocyte effacement and a pathogenicity island of *Escherichia coli* O157:H7 is frequently present in attaching and effacing *E. coli*. Infect Immun. 2003;71(6):3343–8.
18. O'Brien AD, Newland JW, Miller SF, Holmes RK, Smith HW, Formal SB. Shiga-like toxin-converting phages from *Escherichia coli* strains that cause hemorrhagic colitis or infantile diarrhea. Science. 1984;226(4675):694–6.
19. Hayashi T, Makino K, Ohnishi M, Kurokawa K, Ishii K, Yokoyama K, Han CG, Ohtsubo E, Nakayama K, Murata T, Tanaka M, Tobe T, Iida T, Takami H, Honda T, Sasakawa C, Ogasawara N, Yasunaga T, Kuhara S, Shiba T, Hattori M, Shinagawa H. Complete genome sequence of enterohemorrhagic *Escherichia coli* O157:H7 and genomic comparison with a laboratory strain K-12. DNA Res. 2001;8(1):11–22.
20. Perna NT, Plunkett 3rd G, Burland V, Mau B, Glasner JD, Rose DJ, Mayhew GF, Evans PS, Gregor J, Kirkpatrick HA, Posfai G, Hackett J, Klink S, Boutin A, Shao Y, Miller L, Grotbeck EJ, Davis NW, Lim A, Dimalanta ET, Potamousis KD, Apodaca J, Anantharaman TS, Lin J, Yen G, Schwartz DC, Welch RA, Blattner FR. Genome sequence of enterohaemorrhagic *Escherichia coli* O157:H7. Nature. 2001;409(6819):529–33. doi:10.1038/35054089.
21. Muniesa M, Blanco JE, De Simon M, Serra-Moreno R, Blanch AR, Jofre J. Diversity of stx2 converting bacteriophages induced from Shiga-toxin-producing *Escherichia coli* strains isolated from cattle. Microbiology. 2004;150(Pt 9):2959–71. doi:10.1099/mic.0.27188-0.
22. Strockbine NA, Marques LR, Newland JW, Smith HW, Holmes RK, O'Brien AD. Two toxin-converting phages from *Escherichia coli* O157:H7 strain 933 encode antigenically distinct toxins with similar biologic activities. Infect Immun. 1986;53(1):135–40.
23. Willshaw GA, Smith HR, Scotland SM, Field AM, Rowe B. Heterogeneity of *Escherichia coli* phages encoding Vero cytotoxins: comparison of cloned sequences determining VT1 and VT2 and development of specific gene probes. J Gen Microbiol. 1987;133(5):1309–17.
24. Calderwood SB, Auclair F, Donohue-Rolfe A, Keusch GT, Mekalanos JJ. Nucleotide sequence of the Shiga-like toxin genes of *Escherichia coli*. Proc Natl Acad Sci U S A. 1987;84(13):4364–8.
25. Huang A, de Grandis S, Friesen J, Karmali M, Petric M, Congi R, Brunton JL. Cloning and expression of the genes specifying Shiga-like toxin production in *Escherichia coli* H19. J Bacteriol. 1986;166(2):375–9.
26. O'Brien AD, Marques LR, Kerry CF, Newland JW, Holmes RK. Shiga-like toxin converting phage of enterohemorrhagic *Escherichia coli* strain 933. Microb Pathog. 1989;6(5):381–90.
27. Karch H, Meyer T. Single primer pair for amplifying segments of distinct Shiga-like-toxin genes by polymerase chain reaction. J Clin Microbiol. 1989;27(12):2751–7.
28. Willshaw GA, Smith HR, Scotland SM, Rowe B. Cloning of genes determining the production of vero cytotoxin by *Escherichia coli*. J Gen Microbiol. 1985;131(11):3047–53.
29. Persson S, Olsen KE, Ethelberg S, Scheutz F. Subtyping method for *Escherichia coli* shiga toxin (verocytotoxin) 2 variants and correlations to clinical manifestations. J Clin Microbiol. 2007;45(6):2020–4. doi:10.1128/JCM.02591-06.
30. Scheutz F, Teel LD, Beutin L, Pierard D, Buvens G, Karch H, Mellmann A, Caprioli A, Tozzoli R, Morabito S, Strockbine NA, Melton-Celsa AR, Sanchez M, Persson S, O'Brien AD. Multicenter evaluation of a sequence-based protocol for subtyping Shiga toxins and standardizing Stx nomenclature. J Clin Microbiol. 2012;50(9):2951–63. doi:10.1128/JCM.00860-12.
31. Friedrich AW, Bielaszewska M, Zhang WL, Pulz M, Kuczius T, Ammon A, Karch H. *Escherichia coli* harboring Shiga toxin 2 gene variants: frequency and association with clinical symptoms. J Infect Dis. 2002;185(1):74–84. doi:10.1086/338115.
32. McDaniel TK, Jarvis KG, Donnenberg MS, Kaper JB. A genetic locus of enterocyte effacement conserved among diverse enterobacterial pathogens. Proc Natl Acad Sci U S A. 1995;92(5):1664–8.
33. Frankel G, Candy DC, Everest P, Dougan G. Characterization of the C-terminal domains of intimin-like proteins of enteropathogenic and enterohemorrhagic *Escherichia coli*, *Citrobacter freundii*, and *Hafnia alvei*. Infect Immun. 1994;62(5):1835–42.

34. Hyma KE, Lacher DW, Nelson AM, Bumbaugh AC, Janda JM, Strockbine NA, Young VB, Whittam TS. Evolutionary genetics of a new pathogenic Escherichia species: *Escherichia albertii* and related *Shigella boydii* strains. J Bacteriol. 2005;187(2):619–28. doi:10.1128/JB.187.2.619-628.2005.
35. Lacher DW, Steinsland H, Whittam TS. Allelic subtyping of the intimin locus (eae) of pathogenic *Escherichia coli* by fluorescent RFLP. FEMS Microbiol Lett. 2006;261(1):80–7. doi:10.1111/j.1574-6968.2006.00328.x.
36. Schauer DB, Falkow S. The eae gene of Citrobacter freundii biotype 4280 is necessary for colonization in transmissible murine colonic hyperplasia. Infect Immun. 1993;61(11):4654–61.
37. Delahay RM, Frankel G, Knutton S. Intimate interactions of enteropathogenic *Escherichia coli* at the host cell surface. Curr Opin Infect Dis. 2001;14(5):559–65.
38. Frankel G, Phillips AD, Rosenshine I, Dougan G, Kaper JB, Knutton S. Enteropathogenic and enterohaemorrhagic *Escherichia coli*: more subversive elements. Mol Microbiol. 1998;30(5):911–21.
39. Hartland EL, Batchelor M, Delahay RM, Hale C, Matthews S, Dougan G, Knutton S, Connerton I, Frankel G. Binding of intimin from enteropathogenic *Escherichia coli* to Tir and to host cells. Mol Microbiol. 1999;32(1):151–8.
40. Agin TS, Wolf MK. Identification of a family of intimins common to *Escherichia coli* causing attaching-effacing lesions in rabbits, humans, and swine. Infect Immun. 1997;65(1):320–6.
41. Oswald E, Schmidt H, Morabito S, Karch H, Marches O, Caprioli A. Typing of intimin genes in human and animal enterohemorrhagic and enteropathogenic *Escherichia coli*: characterization of a new intimin variant. Infect Immun. 2000;68(1):64–71.
42. Fitzhenry RJ, Pickard DJ, Hartland EL, Reece S, Dougan G, Phillips AD, Frankel G. Intimin type influences the site of human intestinal mucosal colonisation by enterohaemorrhagic *Escherichia coli* O157:H7. Gut. 2002;50(2):180–5.
43. Boerlin P, McEwen SA, Boerlin-Petzold F, Wilson JB, Johnson RP, Gyles CL. Associations between virulence factors of Shiga toxin-producing *Escherichia coli* and disease in humans. J Clin Microbiol. 1999;37(3):497–503.
44. Gyles CL. Shiga toxin-producing *Escherichia coli*: an overview. J Anim Sci. 2007;85 Suppl 13:E45–62. doi:10.2527/jas.2006-508.
45. Wilson JB, Clarke RC, Renwick SA, Rahn K, Johnson RP, Karmali MA, Lior H, Alves D, Gyles CL, Sandhu KS, McEwen SA, Spika JS. Vero cytotoxigenic *Escherichia coli* infection in dairy farm families. J Infect Dis. 1996;174(5):1021–7.
46. Johnson KE, Thorpe CM, Sears CL. The emerging clinical importance of non-O157 Shiga toxin-producing Escherichia coli. Clin Infect Dis. 2006;43(12):1587–95. doi:10.1086/509573.
47. Mellmann A, Bielaszewska M, Köck R, Friedrich AW, Fruth A, Middendorf B, Harmsen D, Schmidt MA, Karch H. Analysis of collection of hemolytic uremic syndrome-associated enterohemorrhagic Escherichia coli. Emerg Infect Dis. 2008;14(8):1287–90. doi:10.3201/eid1408.071082.
48. Newton HJ, Sloan J, Bulach DM, Seemann T, Allison CC, Tauschek M, Robins-Browne RM, Paton JC, Whittam TS, Paton AW, Hartland EL. Shiga toxin-producing Escherichia coli strains negative for locus of enterocyte effacement. Emerg Infect Dis. 2009;15(3):372–80. doi:10.3201/eid1503.080631.
49. Käppeli U1, Hächler H1, Giezendanner N1, Cheasty T2, Stephan R1. Shiga toxin-producing Escherichia coli O157 associated with human infections in Switzerland, 2000-2009. Epidemiol Infect. 2011; 139(7):1097-104. doi:10.1017/S0950268810002190.
50. Watanabe H, Wada A, Inagaki Y, Itoh K, Tamura K. Outbreaks of enterohaemorrhagic *Escherichia coli* O157:H7 infection by two different genotype strains in Japan, 1996. Lancet. 1996;348(9030):831–2.
51. Kudva IT, Evans PS, Perna NT, Barrett TJ, Ausubel FM, Blattner FR, Calderwood SB. Strains of *Escherichia coli* O157:H7 differ primarily by insertions or deletions, not single-nucleotide polymorphisms. J Bacteriol. 2002;184(7):1873–9.
52. Blattner FR, Plunkett 3rd G, Bloch CA, Perna NT, Burland V, Riley M, Collado-Vides J, Glasner JD, Rode CK, Mayhew GF, Gregor J, Davis NW, Kirkpatrick HA, Goeden MA, Rose

DJ, Mau B, Shao Y. The complete genome sequence of *Escherichia coli* K-12. Science. 1997;277(5331):1453–62.

53. Hacker J, Kaper JB. Pathogenicity islands and the evolution of microbes. Annu Rev Microbiol. 2000;54:641–79. doi:10.1146/annurev.micro.54.1.641.
54. Klapproth JM, Scaletsky IC, McNamara BP, Lai LC, Malstrom C, James SP, Donnenberg MS. A large toxin from pathogenic *Escherichia coli* strains that inhibits lymphocyte activation. Infect Immun. 2000;68(4):2148–55.
55. Nicholls L, Grant TH, Robins-Browne RM. Identification of a novel genetic locus that is required for in vitro adhesion of a clinical isolate of enterohaemorrhagic *Escherichia coli* to epithelial cells. Mol Microbiol. 2000;35(2):275–88.
56. Karmali MA. Use of comparative genomics as a tool to assess the clinical and public health significance of emerging Shiga toxin-producing *Escherichia coli* serotypes. Meat Sci. 2005;71(1):62–71. doi:10.1016/j.meatsci.2005.03.001.
57. Franz E, Delaquis P, Morabito S, Beutin L, Gobius K, Rasko DA, Bono J, French N, Osek J, Lindstedt BA, Muniesa M, Manning S, LeJeune J, Callaway T, Beatson S, Eppinger M, Dallman T, Forbes KJ, Aarts H, Pearl DL, Gannon VP, Laing CR, Strachan NJ. Exploiting the explosion of information associated with whole genome sequencing to tackle Shiga toxin-producing *Escherichia coli* (STEC) in global food production systems. Int J Food Microbiol. 2014;187:57–72. doi:10.1016/j.ijfoodmicro.2014.07.002.
58. Touchon M, Hoede C, Tenaillon O, Barbe V, Baeriswyl S, Bidet P, Bingen E, Bonacorsi S, Bouchier C, Bouvet O, Calteau A, Chiapello H, Clermont O, Cruveiller S, Danchin A, Diard M, Dossat C, Karoui ME, Frapy E, Garry L, Ghigo JM, Gilles AM, Johnson J, Le Bouguenec C, Lescat M, Mangenot S, Martinez-Jehanne V, Matic I, Nassif X, Oztas S, Petit MA, Pichon C, Rouy Z, Ruf CS, Schneider D, Tourret J, Vacherie B, Vallenet D, Medigue C, Rocha EP, Denamur E. Organised genome dynamics in the *Escherichia coli* species results in highly diverse adaptive paths. PLoS Genet. 2009;5(1), e1000344. doi:10.1371/journal.pgen.1000344.
59. van Elsas JD, Semenov AV, Costa R, Trevors JT. Survival of *Escherichia coli* in the environment: fundamental and public health aspects. ISME J. 2011;5(2):173–83. doi:10.1038/ismej.2010.80.
60. ECDC. Shiga toxin-producing *E. coli* (STEC): update on outbreak in the EU (27 July 2011, 11:00). European Centre for Disease Prevention and Control, Outbreak Update; 2011. http://ecdceuropaeu/en/activities/sciadvice/_layouts/forms/Review_DispFormaspx?List=a3216f4c-f040-4f51-9f77-a96046dbfd72&ID=602.
61. Karch H, Denamur E, Dobrindt U, Finlay BB, Hengge R, Johannes L, Ron EZ, Tonjum T, Sansonetti PJ, Vicente M. The enemy within us: lessons from the 2011 European *Escherichia coli* O104:H4 outbreak. EMBO Mol Med. 2012;4(9):841–8. doi:10.1002/emmm.201201662.
62. Scheutz F, Nielsen EM, Frimodt-Moller J, Boisen N, Morabito S, Tozzoli R, Nataro JP, Caprioli A. Characteristics of the enteroaggregative Shiga toxin/verotoxin-producing *Escherichia coli* O104:H4 strain causing the outbreak of haemolytic uraemic syndrome in Germany, May to June 2011. Euro Surveill. 2011;16(24).
63. Werber D, King LA, Muller L, Follin P, Buchholz U, Bernard H, Rosner B, Ethelberg S, de Valk H, Hohle M. Associations of age and sex with the clinical outcome and incubation period of Shiga toxin-producing *Escherichia coli* O104:H4 infections, 2011. Am J Epidemiol. 2013;178(6):984–92. doi:10.1093/aje/kwt069.
64. EFSA. Tracing seeds, in particular fenugreek (Trigonella foenum-graecum) seeds, in relation to the Shiga toxin-producing *E. coli* (STEC) O104:H4 2011 outbreaks in Germany and France. Technical report of the European Food Safety Authority; 2011. http://www.efsaeuropaeu/it/supporting/pub/176ehtm.
65. Brzuszkiewicz E, Thurmer A, Schuldes J, Leimbach A, Liesegang H, Meyer FD, Boelter J, Petersen H, Gottschalk G, Daniel R. Genome sequence analyses of two isolates from the recent *Escherichia coli* outbreak in Germany reveal the emergence of a new pathotype: Entero-Aggregative-Haemorrhagic Escherichia coli (EAHEC). Arch Microbiol. 2011;193(12):883–91. doi:10.1007/s00203-011-0725-6.

66. Mellmann A, Harmsen D, Cummings CA, Zentz EB, Leopold SR, Rico A, Prior K, Szczepanowski R, Ji Y, Zhang W, McLaughlin SF, Henkhaus JK, Leopold B, Bielaszewska M, Prager R, Brzoska PM, Moore RL, Guenther S, Rothberg JM, Karch H. Prospective genomic characterization of the German enterohemorrhagic *Escherichia coli* O104:H4 outbreak by rapid next generation sequencing technology. PLoS One. 2011;6(7), e22751. doi:10.1371/journal.pone.0022751.
67. Rasko DA, Webster DR, Sahl JW, Bashir A, Boisen N, Scheutz F, Paxinos EE, Sebra R, Chin CS, Iliopoulos D, Klammer A, Peluso P, Lee L, Kislyuk AO, Bullard J, Kasarskis A, Wang S, Eid J, Rank D, Redman JC, Steyert SR, Frimodt-Moller J, Struve C, Petersen AM, Krogfelt KA, Nataro JP, Schadt EE, Waldor MK. Origins of the *E. coli* strain causing an outbreak of hemolytic-uremic syndrome in Germany. N Engl J Med. 2011;365(8):709–17. doi:10.1056/NEJMoa1106920.
68. Rohde H, Qin J, Cui Y, Li D, Loman NJ, Hentschke M, Chen W, Pu F, Peng Y, Li J, Xi F, Li S, Li Y, Zhang Z, Yang X, Zhao M, Wang P, Guan Y, Cen Z, Zhao X, Christner M, Kobbe R, Loos S, Oh J, Yang L, Danchin A, Gao GF, Song Y, Li Y, Yang H, Wang J, Xu J, Pallen MJ, Wang J, Aepfelbacher M, Yang R, Consortium EcOHGAC-S. Open-source genomic analysis of Shiga-toxin-producing *E. coli* O104:H4. N Engl J Med. 2011;365(8):718–24. doi:10.1056/NEJMoa1107643.
69. EHEC-crowdsourced. *E. coli* O104:H4 genome analysis crowdsourcing. 2011. https://github-com/ehec-outbreak-crowdsourced/BGI-data-analysis/wiki.
70. Beutin L, Hammerl JA, Reetz J, Strauch E. Shiga toxin-producing *Escherichia coli* strains from cattle as a source of the Stx2a bacteriophages present in enteroaggregative *Escherichia coli* O104:H4 strains. Int J Med Microbiol. 2013;303(8):595–602. doi:10.1016/j.ijmm.2013.08.001.
71. Hebbelstrup Jensen B, Olsen KE, Struve C, Krogfelt KA, Petersen AM. Epidemiology and clinical manifestations of enteroaggregative *Escherichia coli*. Clin Microbiol Rev. 2014;27(3):614–30. doi:10.1128/CMR.00112-13.
72. Valat C, Auvray F, Forest K, Metayer V, Gay E, Peytavin de Garam C, Madec JY, Haenni M. Phylogenetic grouping and virulence potential of extended-spectrum-beta-lactamase-producing *Escherichia coli* strains in cattle. Appl Environ Microbiol. 2012;78(13):4677–82. doi:10.1128/AEM.00351-12.
73. Pritchard L, Holden NJ, Bielaszewska M, Karch H, Toth IK. Alignment-free design of highly discriminatory diagnostic primer sets for *Escherichia coli* O104:H4 outbreak strains. PLoS One. 2012;7(4), e34498. doi:10.1371/journal.pone.0034498.
74. Morabito S, Karch H, Mariani-Kurkdjian P, Schmidt H, Minelli F, Bingen E, Caprioli A. Enteroaggregative, Shiga toxin-producing *Escherichia coli* O111:H2 associated with an outbreak of hemolytic-uremic syndrome. J Clin Microbiol. 1998;36(3):840–2.
75. Dallman T, Smith GP, O'Brien B, Chattaway MA, Finlay D, Grant KA, Jenkins C. Characterization of a verocytotoxin-producing enteroaggregative *Escherichia coli* serogroup O111:H21 strain associated with a household outbreak in Northern Ireland. J Clin Microbiol. 2012;50(12):4116–9. doi:10.1128/JCM.02047-12.
76. Tozzoli R, Grande L, Michelacci V, Ranieri P, Maugliani A, Caprioli A, Morabito S. Shiga toxin-converting phages and the emergence of new pathogenic *Escherichia coli*: a world in motion. Front Cell Infect Microbiol. 2014;4:80. doi:10.3389/fcimb.2014.00080.
77. Grande L, Michelacci V, Tozzoli R, Ranieri P, Maugliani A, Caprioli A, Morabito S. Whole genome sequence comparison of vtx2-converting phages from Enteroaggregative Haemorrhagic Escherichia coli strains. BMC Genomics. 2014;15:574. doi:10.1186/1471-2164-15-574.
78. Beutin L, Hammerl JA, Strauch E, Reetz J, Dieckmann R, Kelner-Burgos Y, Martin A, Miko A, Strockbine NA, Lindstedt BA, Horn D, Monse H, Huettel B, Muller I, Stuber K, Reinhardt R. Spread of a distinct Stx2-encoding phage prototype among *Escherichia coli* O104:H4 strains from outbreaks in Germany, Norway, and Georgia. J Virol. 2012;86(19):10444–55. doi:10.1128/JVI.00986-12.
79. Werts C, Michel V, Hofnung M, Charbit A. Adsorption of bacteriophage lambda on the LamB protein of *Escherichia coli* K-12: point mutations in gene J of lambda responsible for extended host range. J Bacteriol. 1994;176(4):941–7.

80. DG-SANTE. Vision paper on the development of data bases for molecular testing of food-borne pathogens in view of outbreak preparedness. European Commission, Directorate General for Health and Food Safety; 2012. http://eceuropaeu/food/food/biosafety/salmonella/docs/vision-paper_enpdf.
81. EFSA. EFSA's 20th Scientific Colloquium on Whole Genome Sequencing of food-borne pathogens for public health protection. European Food Safety Authority, Scientific Colloquium 20, Parma 16–17 Giugno 2014; 2014. http://wwwefsaeuropaeu/it/supporting/pub/743ehtm.
82. Jolley KA, Maiden MC. Using MLST to study bacterial variation: prospects in the genomic era. Future Microbiol. 2014;9(5):623–30. doi:10.2217/fmb.14.24.
83. Kovanen SM, Kivisto RI, Rossi M, Schott T, Karkkainen UM, Tuuminen T, Uksila J, Rautelin H, Hanninen ML. Multilocus sequence typing (MLST) and whole-genome MLST of *Campylobacter jejuni* isolates from human infections in three districts during a seasonal peak in Finland. J Clin Microbiol. 2014;52(12):4147–54. doi:10.1128/JCM.01959-14.
84. Revez J, Zhang J, Schott T, Kivisto R, Rossi M, Hanninen ML. Genomic variation between *Campylobacter jejuni* isolates associated with milk-borne-disease outbreaks. J Clin Microbiol. 2014;52(8):2782–6. doi:10.1128/JCM.00931-14.
85. Bennett JS, Jolley KA, Maiden MC. Genome sequence analyses show that Neisseria oralis is the same species as 'Neisseria mucosa var. heidelbergensis'. Int J Syst Evol Microbiol. 2013;63(Pt 10):3920–6. doi:10.1099/ijs.0.052431-0.
86. Kohl TA, Diel R, Harmsen D, Rothganger J, Walter KM, Merker M, Weniger T, Niemann S. Whole-genome-based *Mycobacterium tuberculosis* surveillance: a standardized, portable, and expandable approach. J Clin Microbiol. 2014;52(7):2479–86. doi:10.1128/JCM.00567-14.
87. Lee Y, Kim BS, Chun J, Yong JH, Lee YS, Yoo JS, Yong D, Hong SG, D'Souza R, Thomson KS, Lee K, Chong Y. Clonality and resistome analysis of KPC-producing *Klebsiella pneumoniae* strain isolated in Korea using whole genome sequencing. Biomed Res Int. 2014;2014:352862. doi:10.1155/2014/352862.
88. Leopold SR, Goering RV, Witten A, Harmsen D, Mellmann A. Bacterial whole-genome sequencing revisited: portable, scalable, and standardized analysis for typing and detection of virulence and antibiotic resistance genes. J Clin Microbiol. 2014;52(7):2365–70. doi:10.1128/JCM.00262-14.
89. Joensen KG, Scheutz F, Lund O, Hasman H, Kaas RS, Nielsen EM, Aarestrup FM. Real-time whole-genome sequencing for routine typing, surveillance, and outbreak detection of verotoxigenic *Escherichia coli*. J Clin Microbiol. 2014;52(5):1501–10. doi:10.1128/JCM.03617-13.
90. Underwood AP, Dallman T, Thomson NR, Williams M, Harker K, Perry N, Adak B, Willshaw G, Cheasty T, Green J, Dougan G, Parkhill J, Wain J. Public health value of next-generation DNA sequencing of enterohemorrhagic *Escherichia coli* isolates from an outbreak. J Clin Microbiol. 2013;51(1):232–7. doi:10.1128/JCM.01696-12.
91. Donnenberg MS. *Escherichia coli*: virulence mechanisms of a versatile pathogen. San Diego: Academic; 2002.
92. Wirth T, Falush D, Lan R, Colles F, Mensa P, Wieler LH, Karch H, Reeves PR, Maiden MC, Ochman H, Achtman M. Sex and virulence in *Escherichia coli*: an evolutionary perspective. Mol Microbiol. 2006;60(5):1136–51. doi:10.1111/j.1365-2958.2006.05172.x.
93. Jenkins C, Dallman TJ, Launders N, Willis C, Byrne L, Jorgensen F, Eppinger M, Adak GK, Aird H, Elviss N, Grant KA, Morgan D, McLauchlin J. Public health investigation of two outbreaks of Shiga toxin-producing *Escherichia coli* O157 associated with consumption of watercress. Appl Environ Microbiol. 2015;81(12):3946–52. doi:10.1128/AEM.04188-14.
94. Dallman TJ, Byrne L, Ashton PM, Cowley LA, Perry NT, Adak G, Petrovska L, Ellis RJ, Elson R, Underwood A, Green J, Hanage WP, Jenkins C, Grant K, Wain J. Whole-genome sequencing for national surveillance of Shiga toxin-producing *Escherichia coli* O157. Clin Infect Dis. 2015;61(3):305–12. doi:10.1093/cid/civ318.
95. Holmes A, Allison L, Ward M, Dallman TJ, CLark R, Fawkes A, Murphy L, Hanson M. The utility of Whole Genome Sequencing of *Escherichia coli* O157 1 for outbreak detection and epidemiological surveillance. J Clin Microbiol. 2015. doi:10.1128/JCM.01066-15.

96. Dallman T, Ashton P, Byrne L, Perry N, Petrovska L, Ellis R, Allison L, Hanson M, Holmes A, Gunn G, Chase-Topping M, Woolhouse M, Grant K, Gally D, Wain J, Jenkins C. Applying phylogenomics to understand the emergence of Shiga Toxin producing *Escherichia coli* O157:H7 strains causing severe human disease in the United Kingdom. Microbial Genomics. 2015. doi:10.1099/mgen.0.000029.
97. Wick LM, Qi W, Lacher DW, Whittam TS. Evolution of genomic content in the stepwise emergence of *Escherichia coli* O157:H7. J Bacteriol. 2005;187(5):1783–91. doi:10.1128/JB.187.5.1783-1791.2005.
98. Dallman TJ, Byrne L, Launders N, Glen K, Grant KA, Jenkins C. The utility and public health implications of PCR and whole genome sequencing for the detection and investigation of an outbreak of Shiga toxin-producing *Escherichia coli* serogroup O26:H11. Epidemiol Infect. 2015;143(8):1672–80. doi:10.1017/S0950268814002696.

Chapter 8
Campylobacter

Noel McCarthy

Introduction

The genus *Campylobacter* from the delta-epsilon group of proteobacteria, are microaerophilic, Gram-negative, flagellate, spiral bacteria. *Campylobacter jejuni* is the leading cause of bacterial food-borne diarrhoeal disease throughout the world. *C. jejuni* is primarily a cause of gastroenteritis with onset 2–5 days following infection [1]. Laboratory confirmed cases in one large study from the United Kingdom (UK) typically reported diarrhoea (95 %), abdominal pain (85 %) and fever (78 %) and less commonly vomiting (35 %) and bloody diarrhoea (27 %), with factors such as young age and large infecting dose associated with more severe disease [2]. Ten percent of these cases were admitted to hospital. This disease spectrum was based on laboratory confirmed cases in a high-income country. Symptoms are likely to be less severe and certainly hospitalisation rates lower among non-laboratory confirmed clinical cases. Extra-intestinal infection is rare [3, 4] but non-infectious extra-intestinal complications may occur in the weeks following infection notably reactive arthritis and the neurological condition of Guillain-Barré syndrome. Reactive arthritis incidence post gastroenteritis has been estimated at between 1 and 5 % [5] and *C. jejuni* infection may be the most common preceding infectious cause [6] of reactive arthritis. The differential risk for Guillain-Barré syndrome by serogroup [7, 8] supports the application of whole genome sequencing (WGS) to characterise strain characteristics associated with complications more fully, with no evidence for or against pathogen subtype predicting other complications. The closely related species *Campylobacter coli* also causes substantial human disease, perhaps 10 % of the total burden of human campylobacteriosis [9] although there is a limited evidence base

N. McCarthy (✉)
Warwick Medical School, University of Warwick, Gibbet Hill Road, Coventry, CV4 7AL, UK
e-mail: n.d.mccarthy@warwick.ac.uk

X. Deng et al. (eds.), *Applied Genomics of Foodborne Pathogens*, Food Microbiology and Food Safety, DOI 10.1007/978-3-319-43751-4_8

comparing illness across species with routine laboratory identification often ending at the genus level. Because isolation procedures are generally optimised for *C. jejuni* and *C. coli* [10] it is likely that other members of the genus contribute a greater proportion of illness than is recognised, at least in some parts of the world [11, 12], with most evidence to date for *Campylobacter upsaliensis* [13, 14] for which the clinical picture may also be more severe with for example bacteraemia occurring more commonly [13] and more recently reports indicating that *Campylobacter concisus* may be a significant cause of human bacterial gastroenteritis [12]. The advent of genome based molecular tools may support improved detection of the wide range of other species in human samples and in foods [15] and clarify which of these are associated with disease in humans. Partial genome sequencing approaches initially applied to *C. jejuni* [16] and *C. coli* [17] were later extended across other species [18]. Although this chapter focuses mainly on *C. jejuni* and *C. coli* the general approaches described are also applicable across other species and are likely to contribute to the identification of other species in potential sources [15] as well as in cases of human infection.

Future Applications of Genomics

The likely areas for the early application of genomics in the pathogens on which this chapter focuses are: (1) enhancement of descriptive epidemiology to refine and answer outstanding questions in this area; (2) attribution of human infection to source; (3) identification and investigation of outbreaks; and (4) studies of particular features such as antimicrobial resistance, virulence and factors affecting food chain survival. *Epidemiology*: a seasonal peak during summer was an early finding in the study of campylobacteriosis [19, 20] and has been described widely since then [21, 22] with a sudden very sharp increase, often doubling the weekly incidence of reported infection over the course of 2 weeks at the same time each year [21] in most temperate countries. Seasonal variation in clonal groups across the seasons [23] suggest that WGS may contribute to improved understanding of the biology of this feature as well as potentially contributing to understanding of other temporal and geographical patterning of human campylobacteriosis that appear to vary by subtype [23]. *Attribution*: identification of the sources of infection of foodborne zoonoses with more than one animal host or environmental reservoir [24] can guide disease control interventions. Although a lot is known about the main sources of human campylobacter infection from a combination of analytical epidemiology, observation of natural experiments, risk assessment and microbial typing much uncertainty still remains on the quantitative contribution of different sources and transmission routes. Contaminated poultry is widely identified as the most important source, followed by infection from ruminant sources and then a wide range of other wild animal and environmental sources contributing less certain numbers of cases [25]. Seven locus multilocus sequence type information has been extensively exploited in attributing *C. jejuni* and *C. coli* [26–28] to the main sources of human infection. There is some evidence from population genetic analyses of multilocus sequence type (MLST) data and WGS analysis that more accurate attribution may be possible using whole genome data [29, 30]. The transmission routes from reservoir sources to humans are unclear with potential for cross contamination in kitchens being important [31, 32] as well as some evidence for environmental transmission from food animals to humans in addition to transmission via contaminated food [33, 34]. Although this evidence shows that food animal strains can be transmitted by routes other than food consumption outdoor exposures are in general more strongly associated with acquisition of environmentally associated subtypes and cases in pet owners have been reported to be associated with subtypes carried by pets [34, 35]. Whether pathogen genomics can contribute to elucidation of different transmission routes is uncertain and depends on the rate of detectable genetic change during transmission. Differential survival of *Campylobacter* subtypes through food processing has been reported [36] suggesting that some information may be available from WGS studies across the food chain but the practical utility of this is unclear. *Outbreak detection*: although pulse field gel electrophoresis has not been effective in identifying outbreaks of *Campylobacter*, in contrast to the PulseNet findings for other bacteria [37, 38], there is some evidence that population based genomic surveillance can identify outbreaks [39]. However this data also suggests that single strain outbreaks may generally be both small and uncommon [39]. *Particular features*: alongside these broader themes extensive application of genomics to pathogens should allow identification of the genetic basis for a wide range of specific features such as antimicrobial resistance and virulence. Here the capacity of WGS to assay both individual genetic determinants of, for example, virulence and the wider evolutionary context of the isolate from analysis across the rest of the genome is particularly important. After a brief review of MLST and WGS applications to date the main body of this chapter considers how WGS may address these four themes.

Applications of Genome Sequencing to *Campylobacter* to Date

The first sequenced genome of *Campylobacter* [40] was followed shortly after by a widely accepted MLST scheme [16]. Together these have provided the basis for demonstrating the value of population genetic approaches to the population level study of *Campylobacter* species compared to typological approaches, allowing for example improved source attribution as noted above. Although there has been very limited published whole genome based work on large sample collections to date, with most work restricted to 7-locus MLST, some insights into both source attribution [30] and outbreak detection [39] have already been demonstrated using WGS. This early work motivates further application of these approaches. Equally importantly this work has both identified the reliability of Illumina sequencing to produce relatively complete high quality coverage of the genome efficiently [39], and highlighted the combination of appropriate sampling frames and analytical approaches that will be required alongside genome sequencing to answer the range of questions that are open to WGS technologies.

Issues in and Approaches to Applying WGS to *Campylobacter* Epidemiology, Attribution, Outbreak Detection and Other Areas

Descriptive Epidemiology of *Campylobacter*

The substantial seasonal peak in humans [21] is a leading unresolved question in the basic descriptive epidemiology of human campylobacteriosis. Complexities to this include some variation in the seasonality when different age groups are considered separately [41]. A parallel seasonal peak in chickens is one possible driver of this [42]. However given, for example, a wider range of subtypes in the chicken *Campylobacter* population and more stability in these over 3 years than among the isolates from humans [43] the seasonal peaking of human infection at this time may involve sources other than foodborne transmission from poultry. Application of MLST has identified the substantial contribution of a single clonal group identified by ST-45 and ST-283 complex to the human summer peak [9, 23], a clonal group with a wide host range [44–46]. However it is not known whether this is due to increased transmission to humans from the same sources as at other times of year or different additional sources of this clonal group during summer months. Findings to date therefore suggest that the additional information accessible via WGS, in combination with well sampled studies, should clarify this question for this clonal group and allow fuller investigation of the wide range of clonal groups which have seasonally varying patterns. Joint analysis of WGS data alongside other epidemiological factors such as age and urban-rural classification of residence of human cases will be required given the association of seasonal patterns with these patient characteristics. In addition the seasonal ecology in natural reservoirs and seasonal variation in transmission between and from them must be mapped out to allow robust conclusions on the processes driving the temporal pattern of human disease overall and among different sections of the human population. The underlying assumption to the contribution of WGS in this is that it will allow the identification of lineages associated with different potential sources of human infection.

Similarly the geographic patterning identified as present, but generally weak, among many host species will be amenable to more refined assessment with WGS data from large well sampled isolate collections [23, 29, 47, 48]. Most sampling to date has been in industrialised countries. Alongside the relatively modest geographic effects observed overall there is evidence that some lineages, not yet widely identified elsewhere, comprise a substantial proportion of human disease in individual countries. This has been observed in both industrialised [49] and less industrialised countries [50]. Studies integrating temporal, geographical and bacterial population genetic data across multiple datasets before WGS data have allowed additional insights including demonstration of the mutual dependence in the relationship between season, geography and bacterial populations [23]. The critical work to allow use of WGS data to support improved understanding of the basic descriptive epidemiology of *Campylobacter* is therefore the assembly of extensively sampled structured datasets, with epidemiological information for each isolate, and the development of systems to share and jointly analyse these data.

Source Attribution

Host association of some genotypes showed that subtype based attribution might be feasible for *Campylobacter* [51]. However, in contrast to for example *Salmonella* [52], no robust methods of indexing associations was described and no method developed to use the observed host association to attribute human campylobacteriosis to source before the advent of MLST. The advent of MLST allowed population genetic analysis of multi-locus genetic information as well as attribution based on a summary type. Critically, this demonstrated that summary type is not the most efficient approach to predicting the source of human infection. Analysis that considered each of the seven MLST loci separately allowed more accurate prediction [29]. Investigation of the basis for this result suggested that it was due to this approach being able to use information generated by lateral gene transfer within different host species [29] as summarised in Fig. 8.1 (from Emerg Infect Dis. 2007;13:267–72).

The resulting inference that the number of loci at which an isolate appears to have imported genes most commonly present in, for example, chicken derived isolates allows to predict an increased probability of origin in that host reservoir has informed the model based analysis attributing human infection to sources that are the current standard in this area [26–28, 53]. Multi-host lineages may be particularly common across farmed animals making host attribution based on imported genes important in monitoring the control of foodborne disease. Although these approaches are informative there is substantial residual uncertainty as to source for individual isolates after MLST based prediction [29]. However the identification that the lateral gene transfer signal indexed by seven gene MLST contributes to source attribution also suggests that more accurate attribution will be possible with large WGS datasets allowing information on the likely origin of a far greater number of laterally transferred genes to be considered and used. This assumes that their behaviour mirrors that of the seven housekeeping loci in the MLST scheme. To date no large whole genome reference datasets have been used to test this prediction. In contrast marked genetic differentiation among, for example, some apparently wild bird restricted lineages [48] supports more accurate attribution even when using a summary measure of MSLT type alone and the improvements that will be possible from fuller WGS information for attribution of these are uncertain.

In addition to this neutral model population genetic approach, whole genome datasets allow investigation of host association based on selection, thus allowing

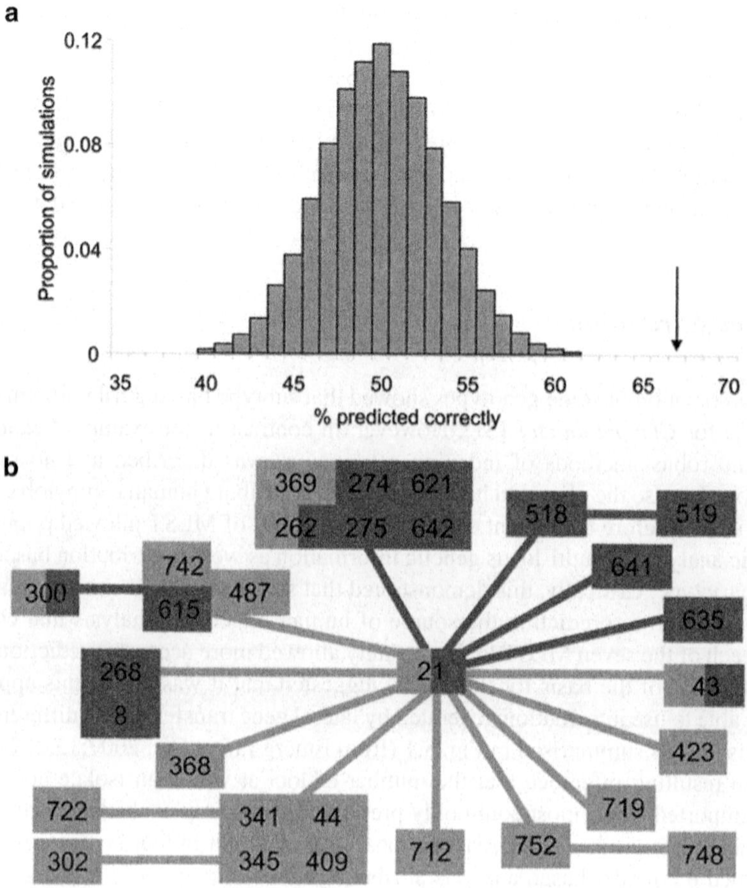

Fig. 8.1 Prediction of source of origin within the sequence type ST-21 complex. (**a**) Observed accuracy of prediction by analysis of imported alleles (*arrow*) compared with distribution of values obtained by permuting host labels so that the alleles varying from central genotype are not informative on host of origin. (**b**) Prediction of origin by using only alleles for which substantial reference information is available. *Light lines* indicate alleles different from ST-21 present mainly in chickens in the reference population (i.e., an allele that would predict chicken origin); *dark lines* indicate alleles present mainly in bovids (i.e., predicts bovid origin). *Light boxes* indicate STs found only in chickens, *dark boxes* indicate STs found only in bovids, and *boxes* with *light* and *dark shading* indicate STs found in bovids and chickens (from Emerg Infect Dis. 2007;13:267–72)

integration of this aspect of biology in population genetic analysis. One study seeking the strongest signal of host association across the genome identified the potential for genes involved in vitamin B5 synthesis to predict host through population genetic analysis. The study confirmed better growth in a vitamin B5 depleted environment for cattle origin isolates than chicken origin and proposed that this may reflect the different diets of farmed poultry where vitamin B5 is likely to be present given a high cereal content in diet, than cattle where it is not [30]. Different *gyrA* alleles, that encode resistance to fluoroquinolone antimicrobials have also been identified as associated with host of origin [54]. Taken together these findings from recombination signal in MLST housekeeping genes, vitamin B5 biosynthesis genes, and allelic forms of *gyrA* show the extent of information that may be present to predict host across a genome, and the different biological processes that may be producing the resulting host prediction signal. Synthesis of these different forms of information, with analyses based on models appropriate to the biology generating each of them is likely to be a central challenge in the use of whole genome sequence data to predict source of human infection even when extensive reference datasets are in place.

Several additional issues may affect accuracy of attribution. WGS may allow more refined insight into geographical structuring of *Campylobacter* populations as noted when considering epidemiology. Although generally relatively weak compared to host association as assayed by MLST geographic associations do appear to exist within host reservoirs [23, 29]. Moreover these differences can cause bias in attribution when the source populations are not geographically matched to those to which cases were exposed [55]. When geographical effects have been clarified adjustment of analyses may be possible to allow some use of the maximum available reference data. In addition to correcting for geographical effects the geography of origin may itself be of importance in disease control. In the same way that a particular association of human disease with one of the three main poultry producers in New Zealand was possible using seven gene MLST analysis [49] attribution to particular national or sub-national origin may be more practicable using WGS and could inform control and monitoring programmes where such geographical information on source is important. Alongside optimising analysis, WGS data may also help to clarify the limits of genome based attribution for some sources. The greater sharing of *Campylobacter* populations between phylogenetically distant food animals [56] in comparison to, for example, wild birds [48] may reflect frequent transmission across host species in the farm setting so that the period of time spent in the most recent host species may not be producing the dominant signal. Reliable identification of the most recent genetic changes may not be possible and indeed may not always be informative of the most recent host. Variation in transmission routes by season and over longer time periods may also produce bias if not considered in analysis. Lastly, the absence of sampling of all possible sources of infection means that it is not possible to attribute appropriately to all sources. The relatively recent identification and sequencing of a vole associated isolate [57] that was very different to populations from other animal hosts emphasises that our environmental sampling remains incomplete. Overall therefore source attribution requires not only the application of WGS to isolates but a mapping out of underlying associations that can confound apparent host association and methodological developments to appropriately incorporate genomic signals of phylogeny, and lateral gene transfer alongside other data supporting prediction of source.

Outbreak Detection and Investigation

The sources of *Campylobacter* outbreaks in the UK and the United States of America (USA), the countries with the most extensive available published data, have moved from being most typically waterborne to foodborne [32, 58–61]. Among foods, milk associated outbreaks have been common and generally associated with unpasteurised milk or failures of pasteurisation, but these have also decreased proportionately over time [32, 59, 62] with poultry consumption becoming more prominent [60]. Outbreaks reported from other industrialised countries have also mainly resulted from contamination of water, milk or chicken and with evidence of cross contamination in some investigations [62]. However, the most striking feature of outbreaks in this pathogen compared to other food borne disease pathogens is how few are detected against a large background of apparently sporadic cases. For example in two studies, UK reviews reported that outbreaks comprised only 0.2 % [59] and 0.4 % of cases [32], respectively. The first population based application of WGS to human surveillance provided further evidence that large single strain outbreaks are uncommon, supporting a view that the lack of detected outbreaks is not just a feature of inadequate surveillance but appears to reflect the occurrence of only few single strain outbreaks, at least at a local level [39]. This analysis also emphasised the capacity for WGS to discriminate between isolates. Across the 1026 loci comparable across all 379 isolates in the study pairwise comparison showed differences at 877 loci on average. In contrast repeated isolations from the same patient varied at between three and 14 genes in analyses using all 1643 loci with information available indicating both the marked diversity of the population and close relationship between isolates that are truly closely epidemiologically related. A similarly close relationship has been confirmed among 20 isolates from different patients in a confirmed outbreak [63]. These initial studies thus suggest that integration of WGS in outbreak surveillance is likely to be effective in outbreak detection if applied but that the yield might be limited which questions the cost-benefit balance of the widespread application to this common pathogen when considering outbreak detection alone. The absence of systematic WGS surveillance opens questions as to the extent to which sentinel surveillance may usefully detect outbreaks as well as monitoring overall sources of population infection. For example the substantial contribution to a large increase in disease burden in New Zealand of a single strain as indexed by seven-locus MLST that was partially linked to a single poultry producer [26] could be interpreted as a large sustained national outbreak. A limited WGS sentinel surveillance scheme would be likely to have detected this early and could have contributed substantially to intelligence to guide and monitor control interventions.

The attribution of a large proportion of apparently sporadic campylobacteriosis to chicken [27, 28, 49] and the findings in New Zealand raise the question of whether widely dispersed outbreaks associated with widely distributed foods might contribute to substantially to the burden of apparently sporadic disease [64]. *Campylobacter*, with a capacity to contaminate food and to then persist but not to multiply [65, 66], in combination with a relatively low infectious dose [67–69], might be expected to be particularly prone to producing outbreaks resulting from contamination high up in the

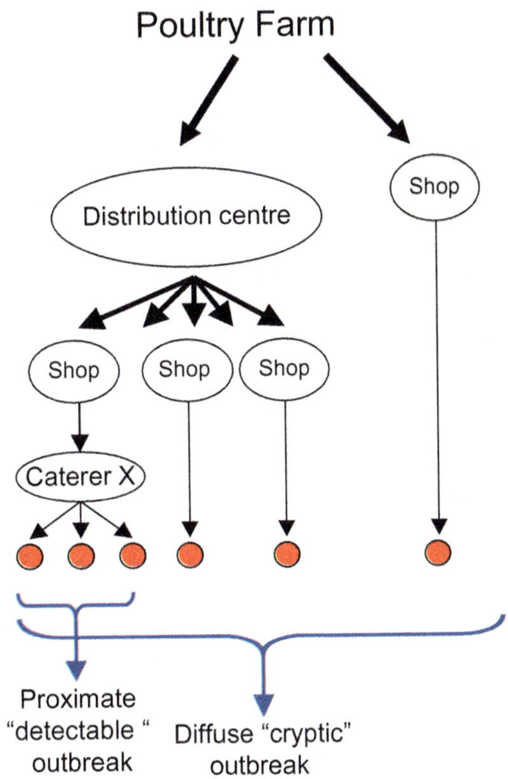

Fig. 8.2 A food chain schematic indicating how contamination high up the food chain could give rise to epidemiologically related cases where this relationship is likely to be missed. These diffuse hidden or "cryptic" outbreaks may include sub-outbreaks that are identifiable such as a function served by Caterer X where food safety failures may lead to some cases clustered identifiably within a subgroup of the population while other infections with the same strain lack such clustering

food chain. Subsequent wide distribution could mean that spatial clustering among cases is unlikely and temporal clustering may be limited (Fig. 8.2). These outbreaks would be difficult to detect by classical epidemiological means but potentially amenable to the integration of WGS even in a sentinel surveillance design [39, 64]. The occurrence of some smaller outbreaks with clustering across classical epidemiological parameters within a larger diffuse outbreak may help detection and investigation when these are correctly linked together (Fig. 8.2) as evidence in the investigation of a widespread atypical foodborne verocytotoxin producing *E. coli* outbreak [70].

The most obvious potential source of such "outbreaks", where several cases are infected from a common source high in the food chain is via contaminated poultry. However, other possible sources could include for example, contaminated incompletely pasteurised milk or milk products [63]. Integrating WGS into systematic or even sentinel surveillance could act as a safeguard against the development of substantial undetected outbreaks far up the food chain. This function of WGS surveillance assumes that outbreaks will be single strain or at least include a dominant single strain or set of strains [63]. The capacity of such human sentinel surveillance to contribute to public health intelligence will be far greater if it is undertaken alongside food source WGS surveillance. This could allow both the prediction of likely host source based on WGS based attribution as described above and possibly direct linkage back to a specific source if host source monitoring is sufficiently intensive allowing identification of epidemiologically related strains on farms, foods and in the faeces of human cases. At current costs such farm to human faeces monitoring is unlikely to be extensive. However, given substantial testing by industry as well as regulatory authorities a collaborative data sharing model may make a partial form of such surveillance possible in some countries in the near future. Alongside extensive sampling requirements such surveillance would pose analytical challenges. Given the diversity of *Campylobacter* genomes, large diverse datasets cannot currently be compared in a way that also allows fine discrimination to confirm that isolates are closely related. This therefore requires some form of hierarchical comparison strategy to effectively link isolates across surveillance systems [39, 71]. Additionally, the detection of outbreaks associated with contamination high in the food chain may be complicated by evolution of pathogens along the food chain. There is no data to support interpretation of the extent to which isolates from a distant contaminated source may nonetheless be very similar or may have undergone relatively substantial evolution. The biology of *Campylobacter* with a lack of growth on foodstuffs might be predicted to produce little genetic change but this remains uncertain in the absence of empirical evidence. Investigation of such outbreaks if detected, where WGS data can be analysed in the context of other epidemiological information, is likely to provide useful estimates of the genomic changes that occur in these contexts. If outbreaks with a structured distribution network, such as that shown in Fig. 8.2 are identified and investigated by methods including WGS it may be possible to identify whether isolates that are epidemiologically closer (e.g. from cases in the sub-outbreak associated with Caterer X) are also detectably clustered within the outbreak as a whole by WGS. Fully addressing these questions will require the development of improved food chain mapping in outbreaks to support the joint analysis of genetic distance and distance in the food chain.

Just as the cost-effectiveness of integrating WGS into the surveillance of *Campylobacter* is uncertain, the investigation of small outbreak signals suggested by clustering of highly similar isolates in genomic space and other dimensions such as time also raises questions of appropriate resource use. The arguments for such investigations include that they may help disease control activities, that the causes of small outbreaks may overlap with the causes of true sporadic cases and therefore the bulk of human infection, and that small outbreaks may be detectable subsets of larger diffuse outbreaks. Against it are the costs and the likelihood that even after full investigation the sources of small outbreaks are often incompletely identified.

Virulence, Antimicrobial Resistance and Other Specific Areas in the Genome Era

Approaches to studying the virulence of pathogens using WGS are developing rapidly. These include methodological approaches to apply genome wide association studies (GWAS) to microbial pathogens [72] and practical applications assaying virulence factors by genomic comparisons across the genus *Campylobacter* [73] focussed on a single clonal complex within *C. jejuni* [74] and even a single strain [75]. Substantively, these genome wide studies of virulence support a model of horizontal gene transfer as the main mechanism for evolution of virulence in *Campylobacter*, or at least those aspects of virulence studied. Methodologically, joint comparison of both highly clonally related and clonally distant strains is a common feature across these examples which mainly focus on genomic comparisons. These analyses may be particularly efficient in detecting virulence evolving by lateral gene transfer while other approaches may be better at identifying other mechanisms. The single strain focussed work also included a multi-omics approach to demonstrate both up and down regulation of a range of metabolic pathways in a virulent strain as indexed by transcriptomic and proteomic analyses [75]. Separate work on this strain has also suggested that methylation may be a factor in it's pathogenic phenotype (an association with abortion in sheep) [76] highlighting that genome methylation, an area where bacterial genome sequencing approaches are increasing in sophistication [77], may contribute to virulence in *Campylobacter* as may post-transcriptional RNA methylation [78]. These early analyses demonstrate the power of both breadth and depth based approaches to assaying virulence factors in the genome and their associated biology. Given the increasing facility of genome sequencing large populations of isolates the critical factors supporting analysis across this breadth of data will be the development and validation of methods and the availability of phenotypic data associated with isolates. Two of the examples cited [74, 75] used comparison with *C. jejuni* subsp *doylei* as a comparator with known high virulence. Although this is currently an efficient pragmatic approach it is clearly limited and can pick out as virulence associated any genes present in a strain under study and this typically virulent clade, whether these genes are individually associated with virulence or not. More systematic collection of phenotypic data across large sampled populations of genome sequenced isolates including phenotypic measures of virulence is essential. The structured sampling and analytical approaches to identifying virulence are likely to also allow the reliable identification of other features such as survival in the food chain and antimicrobial resistance.

Although antimicrobial treatment is not usual in the clinical care of human campylobacteriosis antimicrobial resistance in this pathogen remains an area of interest for several reasons. Firstly, as outlined above resistance associated mutations may contribute to the attribution of strains to host species of origin [54] emphasising that resistance is patterned across the natural host ecology of *Campylobacter*. Secondly, the large numbers of humans infected and sampled and wide range of wild and domesticated animal host species sources of this *Campylobacter* species make it a potential model genus for understanding gene flow between other species and humans and the impact of animal treatment on transmission of antimicrobial resistance elements to humans. Thirdly, resistance elements transmitted to humans and more particularly between other animal species, may transfer to non-*Campylobacter* species of greater clinical or economic importance [79]. Studies using MLST to map the population structure in relation to antimicrobial resistance [79–84] have identified a range of patterns supportive of clonal spread of resistance [82–85] although findings consistent with recurrent *de-novo* evolution and lateral gene transfer are also reported [80, 81] with one study identifying stronger antimicrobial resistance associations with sources of isolation than genotype [86] suggesting that clonal expansion was not a dominant process of spread of resistance in the sample studied. Almost all of these mechanisms are likely to be important given the very different ecologies and antimicrobial pressures experienced by *Campylobacter* populations in different reservoirs and transmission systems. The capacity of WGS to simultaneously assay the genetic patterns of resistance and a detailed population structure, as indexed across the rest of the genome, is likely to be particularly important in understanding the generation, spread, and sometimes loss of antimicrobial resistance in *Campylobacter*. Although there is already substantial evidence for the association of resistance among food animals [84, 85, 87] and some studies demonstrating antimicrobial resistance differences in food animal origin isolates compared to wild animal or environmental isolates [83, 86] the process of generation and transmission of resistance is very incompletely understood. The detailed phylogeny alongside direct genomic evidence for resistance available from WGS is likely to substantially transform this understanding. As already undertaken for other pathogens [88] a process of mapping the genetic basis of resistance against phenotypically demonstrated resistance is a critical early step to fully exploiting WGS to study the pattern and spread of resistance. Additionally, the inferences will depend on the sampling frame generating sequenced isolates. Sampling across human, food chain and wild animal and environmental settings will be needed to map the generation and flow of resistant strains and resistance elements. Studies on bacterial populations with known antibiotic use exposures will be required to extend this work to understand the relationship between these exposures and the pattern of antimicrobial resistance produced, evidence that could guide interventions to reduce resistance.

Conclusion

Although most of the current and potential developments described above require the availability of WGS from cultured isolates the parallel developments, in part arising from genome data, of improved PCR diagnostics [89] may be employed increasingly. The clinical predictive values of such tests when positive is uncertain and susceptible to the effect of frequent carriage of *Campylobacter* not causing disease reducing test specificity [89]. Evidence that even with culture asymptomatic infection and carriage may be common is an important context for interpreting such results. Tests using PCR may extend the detectability of asymptomatic carriage increasing this effect. Quantification or culture confirmation may help interpretation of positive molecular tests for the presence of *Campylobacter* DNA. Additionally confirmation by culture could mitigate the negative impact of the expansion of direct molecular tests in clinical or food chain settings on the availability of isolates for WGS characterisation. Until direct WGS is possible on clinical and food samples, maintenance of culture for either primary identification or confirmation of positive tests will be needed to support the generation of the potential benefits outlined above. In the longer term both metagenomics approaches and individual strain WGS direct from clinical and other samples may supersede these caveats and allow much fuller understanding that currently available genomic approaches.

References

1. Karmali MA, Fleming PC. *Campylobacter* enteritis. Can Med Assoc J. 1979;120(12):1525–32.
2. Gillespie IA, O'Brien SJ, Frost JA, Tam C, Tompkins D, Neal KR, et al. Investigating vomiting and/or bloody diarrhoea in *Campylobacter jejuni* infection. J Med Microbiol. 2006;55(Pt 6):741–6.
3. Skirrow MB, Jones DM, Sutcliffe E, Benjamin J. *Campylobacter* bacteraemia in England and Wales, 1981–91. Epidemiol Infect. 1993;110(3):567–73.
4. Skirrow MB, Blaser MJ. Clinical aspects of *Campylobacter* infection. In: Nachamkin I, Blaser MJ, editors. Campylobacter. Washington, DC: ASM; 2000.
5. Pope JE, Krizova A, Garg AX, Thiessen-Philbrook H, Ouimet JM. *Campylobacter* reactive arthritis: a systematic review. Semin Arthritis Rheum. 2007;37(1):48–55.
6. Soderlin MK, Kautiainen H, Puolakkainen M, Hedman K, Soderlund-Venermo M, Skogh T, et al. Infections preceding early arthritis in southern Sweden: a prospective population-based study. J Rheumatol. 2003;30(3):459–64.
7. Engberg J, Nachamkin I, Fussing V, McKhann GM, Griffin JW, Piffaretti JC, et al. Absence of clonality of *Campylobacter jejuni* in serotypes other than HS:19 associated with Guillain-Barré syndrome and gastroenteritis. J Infect Dis. 2001;184(2):215–20.
8. Lastovica AJ, Goddard EA, Argent AC. Guillain-Barré syndrome in South Africa associated with *Campylobacter jejuni* O:41 strains. J Infect Dis. 1997;176 Suppl 2:S139–43.
9. Cody AJ, McCarthy NM, Wimalarathna HL, Colles FM, Clark L, Bowler ICJW, et al. A longitudinal 6-year study of the molecular epidemiology of clinical campylobacter isolates in Oxfordshire, United Kingdom. J Clin Microbiol. 2012;50(10):3193–201.
10. Lastovica AJ, Le Roux E. Efficient isolation of *Campylobacter upsaliensis* from stools. J Clin Microbiol. 2001;39(11):4222–3.
11. Man SM. The clinical importance of emerging Campylobacter species. Nat Rev Gastroenterol Hepatol. 2011;8(12):669–85.
12. Nielsen HL, Ejlertsen T, Engberg J, Nielsen H. High incidence of Campylobacter concisus in gastroenteritis in North Jutland, Denmark: a population-based study. Clin Microbiol Infect. 2013;19(5):445–50.
13. Lastovica AJ, Le Roux E, Penner JL. "Campylobacter upsaliensis" isolated from blood cultures of pediatric patients. J Clin Microbiol. 1989;27(4):657–9.
14. Patton CM, Shaffer N, Edmonds P, Barrett TJ, Lambert MA, Baker C, et al. Human disease associated with "Campylobacter upsaliensis" (catalase-negative or weakly positive Campylobacter species) in the United States. J Clin Microbiol. 1989;27(1):66–73.
15. Trokhymchuk A, Waldner C, Chaban B, Gow S, Hill JE. Prevalence and diversity of Campylobacter species in Saskatchewan retail ground beef. J Food Prot. 2014;77(12):2106–10.
16. Dingle KE, Colles FM, Wareing DR, Ure R, Fox AJ, Bolton FE, et al. Multilocus sequence typing system for *Campylobacter jejuni*. J Clin Microbiol. 2001;39(1):14–23.
17. Dingle KE, Colles FM, Falush D, Maiden MC. Sequence typing and comparison of population biology of *Campylobacter coli* and *Campylobacter jejuni*. J Clin Microbiol. 2005;43(1):340–7.
18. Miller WG, On SL, Wang G, Fontanoz S, Lastovica AJ, Mandrell RE. Extended multilocus sequence typing system for *Campylobacter coli*, *C. lari*, *C. upsaliensis*, and *C. helveticus*. J Clin Microbiol. 2005;43(5):2315–29.

19. Thompson JS, Cahoon FE, Hodge DS. Rate of *Campylobacter* spp. isolation in three regions of Ontario, Canada, from 1978 to 1985. J Clin Microbiol. 1986;24(5):876–8.
20. Lassen J, Kapperud G. Epidemiological aspects of enteritis due to *Campylobacter* spp. in Norway. J Clin Microbiol. 1984;19(2):153–6.
21. Nylen G, Dunstan F, Palmer SR, Andersson Y, Bager F, Cowden J, et al. The seasonal distribution of *Campylobacter* infection in nine European countries and New Zealand. Epidemiol Infect. 2002;128(3):383–90.
22. Kovats RS, Edwards SJ, Charron D, Cowden J, D'Souza RM, Ebi KL, et al. Climate variability and *Campylobacter* infection: an international study. Int J Biometeorol. 2005;49(4):207–14.
23. McCarthy ND, Gillespie IA, Lawson AJ, Richardson J, Neal KR, Hawtin PR, et al. Molecular epidemiology of human *Campylobacter jejuni* shows association between seasonal and international patterns of disease. Epidemiol Infect. 2012;140:1102–10.
24. Batz MB, Doyle MP, Morris Jr G, Painter J, Singh R, Tauxe RV, et al. Attributing illness to food. Emerg Infect Dis. 2005;11(7):993–9.
25. Neimann J, Engberg J, Molbak K, Wegener HC. A case-control study of risk factors for sporadic *Campylobacter* infections in Denmark. Epidemiol Infect. 2003;130(3):353–66.
26. Mullner P, Marshall JC, Spencer SEF, Noble AD, Shadbolt T, Collins-Emerson JM, et al. Utilizing a combination of molecular and spatial tools to assess the effect of a public health intervention. Prev Vet Med. 2011;102(3):242–53.
27. Sheppard SK, Dallas JF, Strachan NJ, Macrae M, McCarthy ND, Wilson DJ, et al. Campylobacter genotyping to determine the source of human infection. Clin Infect Dis. 2009;48(8):1072–8.
28. Wilson DJ, Gabriel E, Leatherbarrow AJ, Cheesbrough J, Gee S, Bolton E, et al. Tracing the source of campylobacteriosis. PLoS Genet. 2008;4(9), e1000203.
29. McCarthy ND, Colles FM, Dingle KE, Bagnall MC, Manning G, Maiden MC, et al. Host-associated genetic import in *Campylobacter jejuni*. Emerg Infect Dis. 2007;13(2):267–72.
30. Sheppard SK, Didelot X, Meric G, Torralbo A, Jolley KA, Kelly DJ, et al. Genome-wide association study identifies vitamin B5 biosynthesis as a host specificity factor in Campylobacter. Proc Natl Acad Sci U S A. 2013;110(29):11923–7.
31. Mylius SD, Nauta MJ, Havelaar AH. Cross-contamination during food preparation: a mechanistic model applied to chicken-borne *Campylobacter*. Risk Anal. 2007;27(4):803–13.
32. Frost JA, Gillespie IA, O'Brien SJ. Public health implications of *Campylobacter* outbreaks in England and Wales, 1995–9: epidemiological and microbiological investigations. Epidemiol Infect. 2002;128(2):111–8.
33. Friesema IH, Havelaar AH, Westra PP, Wagenaar JA, van Pelt W. Poultry culling and Campylobacteriosis reduction among humans, the Netherlands. Emerg Infect Dis. 2012;18(3):466–8.
34. Mughini Gras L, Smid JH, Wagenaar JA, de Boer AG, Havelaar AH, Friesema IH, et al. Risk factors for campylobacteriosis of chicken, ruminant, and environmental origin: a combined case-control and source attribution analysis. PLoS One. 2012;7(8), e42599.
35. Mughini Gras L, Smid JH, Wagenaar JA, Koene MG, Havelaar AH, Friesema IH, et al. Increased risk for *Campylobacter jejuni* and *C. coli* infection of pet origin in dog owners and evidence for genetic association between strains causing infection in humans and their pets. Epidemiol Infect. 2013;141(12):2526–35.
36. Colles FM, McCarthy ND, Sheppard SK, Layton R, Maiden MC. Comparison of Campylobacter populations isolated from a free-range broiler flock before and after slaughter. Int J Food Microbiol. 2010;137:259–64.
37. Gerner-Smidt P, Hise K, Kincaid J, Hunter S, Rolando S, Hyytia-Trees E, et al. PulseNet USA: a five-year update. Foodborne Pathog Dis. 2006;3(1):9–19.
38. Hedberg CW, Smith KE, Besser JM, Boxrud DJ, Hennessy TW, Bender JB, et al. Limitations of pulsed-field gel electrophoresis for the routine surveillance of *Campylobacter* infections. J Infect Dis. 2001;184(2):242–4.
39. Cody AJ, McCarthy ND, Jansen van Rensburg M, Isinkaye T, Bentley S, Parkhill J, et al. Real-time genomic epidemiology of human Campylobacter isolates using whole genome multilocus sequence typing. J Clin Microbiol. 2013;51:2526–34.

40. Parkhill J, Wren BW, Mungall K, Ketley JM, Churcher C, Basham D, et al. The genome sequence of the food-borne pathogen *Campylobacter jejuni* reveals hypervariable sequences. Nature. 2000;403(6770):665–8.
41. Strachan NJ, Rotariu O, Smith-Palmer A, Cowden J, Sheppard SK, O'Brien SJ, et al. Identifying the seasonal origins of human campylobacteriosis. Epidemiol Infect. 2013;141(6):1267–75.
42. Meldrum RJ, Griffiths JK, Smith RM, Evans MR. The seasonality of human campylobacter infection and *Campylobacter* isolates from fresh, retail chicken in Wales. Epidemiol Infect. 2005;133(1):49–52.
43. Hanninen ML, Perko-Makela P, Pitkala A, Rautelin H. A three-year study of *Campylobacter jejuni* genotypes in humans with domestically acquired infections and in chicken samples from the Helsinki area. J Clin Microbiol. 2000;38(5):1998–2000.
44. French N, Barrigas M, Brown P, Ribiero P, Williams N, Leatherbarrow H, et al. Spatial epidemiology and natural population structure of *Campylobacter jejuni* colonizing a farmland ecosystem. Environ Microbiol. 2005;7(8):1116–26.
45. Colles FM, Dingle KE, Cody AJ, Maiden MC. Comparison of *Campylobacter* populations in wild geese with those in starlings and free-range poultry on the same farm. Appl Environ Microbiol. 2008;74(11):3583–90.
46. Colles FM, Ali JS, Sheppard SK, McCarthy ND, Maiden MCJ. Campylobacter populations in wild and domesticated Mallard ducks (*Anas platyrhynchos*). Environ Microbiol Rep. 2011;3(5):574–80.
47. Sheppard SK, Colles F, Richardson J, Cody AJ, Elson R, Lawson A, et al. Host association of Campylobacter genotypes transcends geographic variation. Appl Environ Microbiol. 2010;76(15):5269–77.
48. Griekspoor P, Colles FM, McCarthy ND, Hansbro PM, Ashhurst-Smith C, Olsen B, et al. Marked host specificity and lack of phylogeographic population structure of *Campylobacter jejuni* in wild birds. Mol Ecol. 2013;22(5):1463–72.
49. Mullner P, Spencer SE, Wilson DJ, Jones G, Noble AD, Midwinter AC, et al. Assigning the source of human campylobacteriosis in New Zealand: a comparative genetic and epidemiological approach. Infect Genet Evol. 2009;9:1311–9.
50. Duim B, Godschalk PC, van den Braak N, Dingle KE, Dijkstra JR, Leyde E, et al. Molecular evidence for dissemination of unique *Campylobacter jejuni* clones in Curacao, Netherlands Antilles. J Clin Microbiol. 2003;41(12):5593–7.
51. Dingle KE, Colles FM, Ure R, Wagenaar JA, Duim B, Bolton FJ, et al. Molecular characterization of *Campylobacter jejuni* clones: a basis for epidemiologic investigation. Emerg Infect Dis. 2002;8(9):949–55.
52. Hald T, Lo Fo Wong DM, Aarestrup FM. The attribution of human infections with antimicrobial resistant *Salmonella* bacteria in Denmark to sources of animal origin. Foodborne Pathog Dis. 2007;4(3):313–26.
53. Strachan NJC, Gormley FJ, Rotariu O, Ogden ID, Miller G, Dunn GM, et al. Attribution of Campylobacter infections in northeast Scotland to specific sources by use of multilocus sequence typing. J Infect Dis. 2009;199(8):1205–8.
54. Ragimbeau C, Colin S, Devaux A, Decruyenaere F, Cauchie HM, Losch S, et al. Investigating the host specificity of *Campylobacter jejuni* and *Campylobacter coli* by sequencing gyrase subunit A. BMC Microbiol. 2014;14:205.
55. Smid JH, Mughini Gras L, de Boer AG, French NP, Havelaar AH, Wagenaar JA, et al. Practicalities of using non-local or non-recent multilocus sequence typing data for source attribution in space and time of human campylobacteriosis. PLoS One. 2013;8(2), e55029.
56. Sheppard SK, Colles FM, McCarthy ND, Strachan NJC, Ogden ID, Forbes KJ, et al. Niche segregation and genetic structure of *Campylobacter jejuni* populations from wild and agricultural host species. Mol Ecol. 2011;20(16):3484–90.
57. Hepworth PJ, Ashelford KE, Hinds J, Gould KA, Witney AA, Williams NJ, et al. Genomic variations define divergence of water/wildlife-associated *Campylobacter jejuni* niche specialists from common clonal complexes. Environ Microbiol. 2011;13(6):1549–60.

58. Blaser MJ. Epidemiologic and clinical features of *Campylobacter jejuni* infections. J Infect Dis. 1997;176 Suppl 2:S103–5.
59. Pebody RG, Ryan MJ, Wall PG. Outbreaks of *Campylobacter* infection: rare events for a common pathogen. Commun Dis Rep CDR Rev. 1997;7(3):R33–7.
60. Little CL, Gormley FJ, Rawal N, Richardson JF. A recipe for disaster: outbreaks of campylobacteriosis associated with poultry liver pâté in England and Wales. Epidemiol Infect. 2010;138:1691–4.
61. Tauxe RV, Hargrett-Bean N, Patton CM, Wachsmuth IK. *Campylobacter* isolates in the United States, 1982–1986. MMWR CDC Surveill Summ. 1988;37(2):1–13.
62. Friedman CR, Neimann J, Wegener HC, Tauxe RV. Epidemiology of *Campylobacter jejuni* infection in the United States and other industrialized nations. In: Nachamkin I, Blaser MJ, editors. Campylobacter. 2nd ed. Washington, DC: ASM Press; 2000. p. 121–38.
63. Fernandes AM, Balasegaram S, Willis C, Wimalarathna H, Maiden MC, McCarthy ND. Partial failure of milk pasteurisation as a risk for the transmission of Campylobacter from cattle to humans. Clin Infect Dis. 2015;ePub ahead of print. Epub: 13 June 2015.
64. Tauxe RV. Molecular subtyping and the transformation of public health. Foodborne Pathog Dis. 2006;3(1):4–8.
65. Alter T, Scherer K. Stress response of *Campylobacter* spp. and its role in food processing. J Vet Med B Infect Dis Vet Public Health. 2006;53(8):351–7.
66. Park SF. The physiology of *Campylobacter* species and its relevance to their role as foodborne pathogens. Int J Food Microbiol. 2002;74(3):177–88.
67. Black RE, Levine MM, Clements ML, Hughes TP, Blaser MJ. Experimental *Campylobacter jejuni* infection in humans. J Infect Dis. 1988;157(3):472–9.
68. Nauta MJ, Jacobs-Reitsma WF, Havelaar AH. A risk assessment model for *Campylobacter* in broiler meat. Risk Anal. 2007;27(4):845–61.
69. Robinson DA. Infective dose of *Campylobacter jejuni* in milk. Br Med J (Clin Res Ed). 1981;282(6276):1584.
70. Werber D, King LA, Muller L, Follin P, Buchholz U, Bernard H, et al. Associations of age and sex with the clinical outcome and incubation period of Shiga toxin-producing *Escherichia coli* O104:H4 infections, 2011. Am J Epidemiol. 2013;178(6):984–92.
71. Maiden MC, van Rensburg MJ, Bray JE, Earle SG, Ford SA, Jolley KA, et al. MLST revisited: the gene-by-gene approach to bacterial genomics. Nat Rev Microbiol. 2013;11(10):728–36.
72. Farhat MR, Shapiro BJ, Sheppard SK, Colijn C, Murray M. A phylogeny-based sampling strategy and power calculator informs genome-wide associations study design for microbial pathogens. Genome Med. 2014;6(11):101.
73. Iraola G, Perez R, Naya H, Paolicchi F, Pastor E, Valenzuela S, et al. Genomic evidence for the emergence and evolution of pathogenicity and niche preferences in the genus Campylobacter. Genome Biol Evol. 2014;6(9):2392–405.
74. Kivisto RI, Kovanen S, Skarp-de Haan A, Schott T, Rahkio M, Rossi M, et al. Evolution and comparative genomics of *Campylobacter jejuni* ST-677 clonal complex. Genome Biol Evol. 2014;6(9):2424–38.
75. Wu Z, Sahin O, Shen Z, Liu P, Miller WG, Zhang Q. Multi-omics approaches to deciphering a hypervirulent strain of *Campylobacter jejuni*. Genome Biol Evol. 2013;5(11):2217–30.
76. Mou KT, Muppirala UK, Severin AJ, Clark TA, Boitano M, Plummer PJ. A comparative analysis of methylome profiles of *Campylobacter jejuni* sheep abortion isolate and gastroenteric strains using PacBio data. Front Microbiol. 2014;5:782.
77. Beaulaurier J, Zhang XS, Zhu S, Sebra R, Rosenbluh C, Deikus G, et al. Single molecule-level detection and long read-based phasing of epigenetic variations in bacterial methylomes. Nat Commun. 2015;6:7438.
78. Salamaszynska-Guz A, Taciak B, Kwiatek A, Klimuszko D. The Cj0588 protein is a *Campylobacter jejuni* RNA methyltransferase. Biochem Biophys Res Commun. 2014;448(3):298–302.
79. Frye JG, Lindsey RL, Meinersmann RJ, Berrang ME, Jackson CR, Englen MD, et al. Related antimicrobial resistance genes detected in different bacterial species co-isolated from swine fecal samples. Foodborne Pathog Dis. 2011;8(6):663–79.

80. Thakur S, Gebreyes WA. *Campylobacter coli* in swine production: antimicrobial resistance mechanisms and molecular epidemiology. J Clin Microbiol. 2005;43(11):5705–14.
81. Wang X, Zhao S, Harbottle H, Tran T, Blickenstaff K, Abbott J, et al. Antimicrobial resistance and molecular subtyping of *Campylobacter jejuni* and *Campylobacter coli* from retail meats. J Food Prot. 2011;74(4):616–21.
82. Guyard-Nicodeme M, Rivoal K, Houard E, Rose V, Quesne S, Mourand G, et al. Prevalence and characterization of *Campylobacter jejuni* from chicken meat sold in French retail outlets. Int J Food Microbiol. 2015;203:8–14.
83. Olkkola S, Nykasenoja S, Raulo S, Llarena AK, Kovanen S, Kivisto R, et al. Antimicrobial resistance and multilocus sequence types of Finnish *Campylobacter jejuni* isolates from multiple sources. Zoonoses Public Health. 2015;63:10–9.
84. Wimalarathna HM, Richardson JF, Lawson AJ, Elson R, Meldrum R, Little CL, et al. Widespread acquisition of antimicrobial resistance among Campylobacter isolates from UK retail poultry and evidence for clonal expansion of resistant lineages. BMC Microbiol. 2013;13:160.
85. D'Lima CB, Miller WG, Mandrell RE, Wright SL, Siletzky RM, Carver DK, et al. Clonal population structure and specific genotypes of multidrug resistant *Campylobacter coli* from Turkeys. Appl Environ Microbiol. 2007;73:2156–64.
86. Levesque S, Frost E, Arbeit RD, Michaud S. Multilocus sequence typing of *Campylobacter jejuni* isolates from humans, chickens, raw milk, and environmental water in Quebec, Canada. J Clin Microbiol. 2008;46(10):3404–11.
87. Habib I, Miller WG, Uyttendaele M, Houf K, De Zutter L. Clonal population structure and antimicrobial resistance of *Campylobacter jejuni* in chicken meat from Belgium. Appl Environ Microbiol. 2009;75(13):4264–72.
88. Stoesser N, Batty EM, Eyre DW, Morgan M, Wyllie DH, Del Ojo EC, et al. Predicting antimicrobial susceptibilities for *Escherichia coli* and *Klebsiella pneumoniae* isolates using whole genomic sequence data. J Antimicrob Chemother. 2013;68(10):2234–44.
89. Frickmann H, Schwarz NG, Rakotozandrindrainy R, May J, Hagen RM. PCR for enteric pathogens in high-prevalence settings. What does a positive signal tell us? Infect Dis (Lond). 2015;47(7):491–8.

Chapter 9
Genomics and Foodborne Viral Infections

Saskia L. Smits and Marion P.G. Koopmans

Background

Foodborne illness or disease remains a major public health problem globally with substantial economic impact. It results from the consumption of contaminated food or water containing pathogenic bacteria, viruses, or parasites as well as chemical or natural toxins. Acute gastroenteritis is the most common clinical manifestation of foodborne disease, and diarrhoea, characterized by frequent loose or liquid bowel movements, is a common cause of death in developing countries and the second most common cause of morbidity and mortality in young infants worldwide with up to 0.8–1.5 million deaths each year [1–7]. High population density, limited access to clean water, frequent flooding and poor sanitation render surface water bodies in developing countries particularly vulnerable to faecal contamination, leading to a high prevalence of diarrhoeal diseases in both children and adults when untreated water is used for food preparation or drinking. In industrialized countries, where sanitation is widely available, access to safe water is high and personal and domestic hygiene is relatively good, diarrhoeal diseases remain a significant cause of morbidity among all age groups. In the majority of cases, symptoms are brief, and patients do not require medical attention. Though typically self-limited, infectious

S.L. Smits
Department of Viroscience, Erasmus Medical Center, P.O. Box 2040, 3000 CA Rotterdam, The Netherlands
e-mail: s.smits@erasmusmc.nl

M.P.G. Koopmans (✉)
Department of Viroscience, Erasmus Medical Center, P.O. Box 2040, 3000 CA Rotterdam, The Netherlands

Virology Division, Centre for Infectious Diseases Research, Diagnostics and Screening, National Institute for Public Health and the Environment, 3720 BA Bilthoven, The Netherlands
e-mail: m.koopmans@erasmusmc.nl

X. Deng et al. (eds.), *Applied Genomics of Foodborne Pathogens*, Food Microbiology and Food Safety, DOI 10.1007/978-3-319-43751-4_9

diarrhoea episodes result in millions of physician visits annually. A range of pathogens has been associated with foodborne illness, but a handful of organisms cause the majority of acute gastroenteritis cases [8, 9]. It was not until 1972 that viruses were implicated as aetiological agents in diarrhoea; Norwalk virus was identified in the faeces of patients with diarrhoea, followed by rotaviruses in 1973, and enteric adeno- and astroviruses in 1975 [10–13].

Foodborne viral transmission can occur by consumption of food handled by infected food handlers, by contamination of food during the production process (for instance through contaminated water), or by consumption of products of animal origin harbouring a zoonotic virus (Fig. 9.1). Food handler–associated foodborne illness results from the manual preparation of food by an infected food handler shedding viruses, usually resulting in limited outbreaks [14], although their size may be substantial depending on the nature of the contamination. A problem is that food-handlers may transmit viruses before showing symptoms, or have asymptomatic infections [15, 16]. Food contamination can also occur during primary production, as has been observed in particular in fresh produce such as berries and green onions, or bivalve filter-feeding shellfish. Here the nature of contamination is dependent on location of the production area and nature of sewage contamination. In contrast to food handler–associated contamination, production process contamination events may involve multiple pathogens present in sewage, including animal viruses [17–23]. Zoonotic foodborne infection occurs when meat, organs, or other products from an infected animal are consumed. For viruses, this may be the least common mode of transmission, although the potential for such transmission is a cause for concern with every emerging disease outbreak.

Foodborne pathogens share the mode of transmission (fecal-oral) and their ability to infect hosts following oral inoculation. Symptoms may arise from replication and

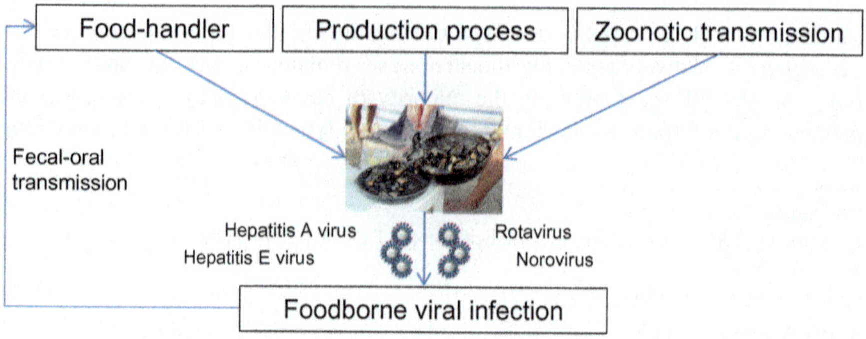

Fig. 9.1 Foodborne viral illness. Foodborne viral transmission can occur by consumption of food handled by infected food handlers, by contamination of food during the production process (for instance through contaminated water), or by consumption of products of animal origin harbouring a zoonotic virus. Foodborne pathogens share the mode of transmission (fecal-oral) and their ability to infect hosts following oral inoculation. Symptoms may arise from replication and the ensuing damage and inflammatory responses in the intestinal tract, but also from generalised infection as observed for instance for orally transmitted hepatitis viruses

the ensuing damage and inflammatory responses in the intestinal tract, but also from generalised infection as observed for instance for orally transmitted hepatitis viruses (hepatitis A and E), or neurotropic enteroviruses. The greatest burden of foodborne viral disease has been attributed to noroviruses and hepatitis A (Fig. 9.1). In addition to these endemic pathogens, the potential for foodborne transmission is a key question in every emerging disease outbreak. In fact, zoonotic emerging infections can be introduced into the population through food preparation or consumption, although these risks are minimal with proper food preparation. Commonly studied in relation to food are viruses from the families Picornaviridae (polio-, entero-, coxsackie-, echo-, and hepatitis A viruses), Reoviridae (rotaviruses), Adenoviridae (adenoviruses 40, 41 primarily), Caliciviridae (noro- and sapoviruses), Hepeviridae (hepatitis E virus), and Astroviridae (Mamastroviruses). They replicate initially in the intestinal tract, are environmentally stable, are shed in high numbers in the faeces of infected individuals with up to 10^{11} virus particles per gram of stool being documented, and are highly infectious with only 10–100 viral particles required for transmission [24, 25]. There is no systematic surveillance for foodborne viral diseases, despite the high burden of disease estimates from some countries [8, 26, 27]. The Food epidemiology reference group (FERG) of the World Health Organisation is currently preparing a global burden of foodborne disease estimate, but the underlying systematic reviews have already signalled large data gaps, particularly from resource limited regions [28].

A combination of factors is responsible for the lack of knowledge of the true incidence of foodborne viral illness. A case of foodborne illness is only identified when a patient falls ill, seeks medical help and undergoes diagnostic testing which leads to identification of the aetiological agent. For some pathogens with long incubation periods (e.g. hepatitis A and E), even when diagnosed, identification of a food source may be extremely difficult due to the long delay between exposure and symptom onset. A third factor compromising the ability to detect foodborne disease is the high rate of asymptomatic infections associated with some pathogens. Therefore, linked cases are difficult to detect. These challenges in diagnosis of foodborne diseases are illustrated by the fact that "unrecognized agents" account for up to 81 % of all U.S. foodborne illnesses and hospitalizations and 64 % of deaths [8, 27, 29]. Rapid population growth and urbanization, deforestation, invasion of previously pristine habitats for agriculture, and increasing demand for animal protein all likely contribute to increased emergence of novel infectious disease threats, while climate change and the increasing global connectedness and mobility facilitate their global spread [30]. Consequently, the pattern of disease outbreaks has changed, from localized clusters of disease in confined populations to dispersed outbreaks with excellent opportunity for further transmission. Similarly, a transition is observed from localized foodborne epidemics to diffuse international foodborne outbreaks due to globalization of the food market [31]. The foodborne nature is often disguised by person-to-person transmission after the initial infection(s) because of the highly infectious nature of most foodborne viral pathogens. Some of these viruses are of major public health concern amongst others because of their food- or waterborne nature, low infectious dose required for infection and serious health-related implications and associated costs.

As in every disease outbreak, including foodborne viral disease outbreaks, the following are some of the most urgent questions to answer: Is the group of ill persons normal for the time of year and/or geographic area or is something extraordinary occurring? If so, which pathogen(s) is causing the disease? Who gets infected? How do people get infected? What is the source of infection? What are transmission routes? How can infection be prevented, treated and/or contained? An integrated multidisciplinary approach utilizing expertise in several areas will be required to understand the dynamics of foodborne viral infection and to mitigate potential effects of future threats. Major challenges regarding recognizing, detecting, characterizing, and effectively responding to foodborne viral threats to health exist, which will be outlined in this chapter, with a focus on how genomics-based tools are a potential candidate to respond to some of these challenges in the field of foodborne viruses.

Foodborne Viruses: What Is Known

Viruses pose a substantial global health burden to humans and the list of human viral infections is ever-changing and continually growing [32]. Mortality in humans from recently emerged viral diseases ranges from a few hundred in the case of severe acute respiratory syndrome (SARS) coronavirus to millions of people from acquired immunodeficiency syndrome (AIDS), caused by human immunodeficiency virus (HIV). We are continuously facing novel pathogens, most of which are zoonotic or originated as zoonoses before adapting to humans [33–35], a proportion of which are likely transmitted via food and/or water [30]. Breakthroughs in the field of metagenomics have had far-reaching effects on the identification and characterization of newly emerging viral pathogens and on the recognition that a growing number of diseases that were once attributed to unknown causes are actually directly or indirectly caused by viral agents [32]. Many previously unknown viruses have been characterized in human stool in recent years including sali-, cosa-, bufa-, picobirna-, reco-, anello-, hepatitis E, astro-, and polyomaviruses of which the clinical disease spectrum, route of transmission, and foodborne nature remains to be elucidated [36–43]. For some of the "older" viruses, such as norovirus and hepatitis A virus, the foodborne risk of transmission is clearly recognized, for others such as adeno- and astroviruses the picture is less clear.

Rotavirus

Although rotaviruses are not generally considered primary foodborne pathogens, because person-to-person transmission seems to be the main route of transmission in developed countries, contaminated water sources are considered to be an important source of rotavirus transmission in developing countries [44]. Rotaviruses are non-enveloped double-stranded segmented RNA viruses from the family *Reoviridae*. The genus Rotavirus contains eight species numbered A–H of which A–C are encountered in humans [45]. Rotavirus A infection is the most common cause of severe gastroenteritis in infants and young children worldwide. Rotavirus B has been found mainly in adults with diarrhoea in China, Bangladesh and India. The viral nucleocapsid outermost layer contains two structural proteins VP4 and VP7 that define the serotype of the virus and are considered critical in vaccine development; more than 40 serotypes have been identified [46]. By the age of five nearly every child has been infected with rotavirus A at least once, the majority of which is anticipated to be symptomatic. The spectrum of rotavirus A disease ranges from mild watery diarrhoea to severe diarrhoea with vomiting and moderate fever. Infection can result in death due to dehydration and electrolyte imbalance that is profuse and life threatening amongst others due to the action of a unique virus encoded enterotoxin NSP4 [47]. The severe impact is primarily observed in young children <2 years of age and can be treated by oral rehydration therapy. Symptoms generally resolve within 3–7 days. Subsequent infections occur from birth to old age but natural immunity renders most of these infections asymptomatic. Rotavirus A is shed in high concentrations in the stool of infected persons and is transmitted via the oral-faecal route with <100 virus particles being sufficient for transmission [45, 48, 49]. Infections occur mainly in late winter or early spring in Europe and colder/drier times of the year in the tropics [50–52]. Rotavirus A vaccines were introduced in 2006, but prior to vaccination policies, rotaviruses caused ~3 million disease episodes per annum in the USA, requiring 500,000 visits to physicians and 60,000 hospitalisations, leading to 20–40 deaths [45, 53–57]. Similar observations were done in Europe [58, 59]. In developing countries rotavirus A infections cause millions of diarrhoea cases, almost two million hospitalizations and an estimated 453,000 infections result in the death of a child younger than 5 years of age annually worldwide [6, 44, 60]. The introduction of proper hygienic measures, clean drinking water, oral rehydration therapy and rotavirus A vaccines reduced disease burden in both developed and developing countries [45].

Norovirus

Noroviruses are positive-stranded RNA viruses belonging to the family *Caliciviridae*. The genus Norovirus is divided into seven genogroups (GI-GVII) that are further subdivided into numerous genotypes [61]. The GI, II, and IV are capable of infecting humans [62], and GII.4 has been associated with the majority of global outbreaks since the mid-1990s. The other genogroups have not been detected in humans, but systematic studies evaluating their role are lacking. Norovirus infections are a leading cause of gastroenteritis outbreaks among all age groups and are transmitted directly from person to person and indirectly via contaminated water and food [63, 64]. They are extremely contagious requiring low viral loads for transmission and are common in closed settings such as healthcare facilities, cruise ships, and nursing homes [24]. The infection can cause nausea, vomiting, watery diarrhoea and abdominal pain, although asymptomatic infections are common [48]. The disease is usually self-limiting, and severe illness is rare in developed countries. Ahmed and coworkers noted a gradient of decreasing prevalence from community to outpatient to inpatient groups, which supports the notion that norovirus is a more common cause of mild acute gastroenteritis [28], although in the USA norovirus infections result in ~70,000 hospitalizations and 800 deaths yearly [65–67]. In developing countries, noroviruses are estimated to cause more than 200,000 deaths annually among children younger than 5 years of age, and it is predicted that these viruses will become the predominant cause of diarrhoea in all age groups worldwide once rotavirus infection is controlled through vaccination [68–71]. The economic impact of foodborne related norovirus gastroenteritis outbreaks is high with an estimated $2 billion healthcare related costs in the USA alone [72].

Hepatitis A Virus

Hepatitis A is a liver disease caused by hepatitis A virus, a non-enveloped positive-stranded RNA virus belonging to the family *Picornaviridae*. Humans are the only naturally known reservoir for hepatitis A viruses and ~5 % of foodborne viral disease is attributed to hepatitis A virus infection [29]. The virus is spread via the faecal-oral route and the disease is closely associated with inadequate sanitation, poor personal hygiene, and limited access to clean water [73–75]. In developing countries, most children are infected with hepatitis A virus by the age of 10 years and the disease is usually asymptomatic in this age group. Epidemics in these countries are practically non-existent as older children and adults are immune to reinfection. In countries with improved sanitary conditions and transitional economies, the rate of infection in young children is lower, resulting in a higher susceptibility of older children and adults and larger outbreaks of disease. The incubation period is 14–28 days and symptoms range from mild to severe, and can include fever, malaise, loss of appetite, diarrhoea, nausea, abdominal discomfort, dark-coloured urine and jaundice which last for up to 8 weeks. Some 10–15 % of people experience a recurrence of symptoms during the 6 months after the initial infection and fulminant hepatitis and acute liver failure occurs although rarely and is most common in the elderly [4, 76]. In developed countries, hepatitis A infection is uncommon and predominantly associated with high-risk groups, such as people travelling to areas of high endemicity. Hepatitis A viruses are stable in the environment and can resist food-production processes routinely used to inactivate and/or control bacterial pathogens. Seroprevalence data indicate tens of millions infections yearly and an estimated 1.4 million clinical cases occur yearly worldwide which have significant social and economic impact [77]. Improved sanitation, food safety and vaccination are the most effective ways to prevent hepatitis A virus infection [75].

Hepatitis E Virus

Hepatitis E virus is a positive-stranded RNA virus with a genome of ~7.2 kb belonging to the family *Hepeviridae*. Four major genotypes are discerned and novel lineages of hepatitis E viruses have been identified in rabbits, rats, wild boar, ferrets and possibly foxes more recently [78–84]. Different genotypes of hepatitis E virus determine differences in epidemiology; genotype 1 is usually seen in developing countries and causes community-level outbreaks while genotype 3 is usually seen in developed countries and rarely causes outbreaks. Hepatitis E virus is transmitted via the faecal-oral route primarily via faecal contamination of water supplies, shellfish and contaminated animal meat, and possibly through zoonosis from pigs. Human-to-human transmission of the virus is rare. Outbreaks and sporadic cases occur worldwide; the virus is most prevalent in East and South Asia and endemic in Asia, Africa and Mexico [85]. An estimated 20 million hepatitis E infections occur worldwide yearly, which are usually self-limited and resolve within 4–6 weeks. Over three million cases of acute fulminant hepatitis E however occur resulting in over 50,000 deaths [4, 86]. Infection with hepatitis E virus is frequent in children in developing countries, but the disease is mostly asymptomatic or causes a very mild illness without jaundice. It causes acute sporadic and epidemic viral hepatitis most commonly in young adults aged 15–40 years with symptoms including jaundice, anorexia, hepatomegaly, abdominal pain and tenderness, nausea, vomiting, and fever that last for up to 2 weeks. A unique disease profile has been observed in pregnant women, where infections with HEV often result in fulminant liver failure, stillbirth and death in 25 % of cases. Treatment and vaccines are unavailable, but currently in development [87].

Enteric Adenovirus

Adenoviruses (Family *Adenoviridae*) are non-enveloped single-stranded DNA viruses with a genome of ~26–48 kb. Adenoviruses infecting humans belong to the genus Mastadenovirus and over 50 serotypes are differentiated based on neutralization assays. Adenoviruses are highly stable in the environment and are thought to spread via respiratory droplets and the faecal-oral route. Adenovirus infections are usually subclinical but certain types are associated with disease which can range from respiratory disease, keratoconjunctivitis, to gastrointestinal disease [88]. Especially human adenoviruses F types 40 and 41 are associated with diarrhoea in young children with acute gastroenteritis and are another major cause of infantile viral diarrhoea in developing countries, following rota- and noroviruses. Symptoms include watery diarrhoea with mucus, fever, dehydration, abdominal pain, and vomiting lasting for 3–11 days [89].

Astrovirus

Astroviruses (Family *Astroviridae*) are non-enveloped positive-stranded RNA viruses with a genome of ~7–8 kb. Classically, eight human serotypes have been described, although since 2008 a large increase in detection of different human astrovirus variants is observed. Human astroviruses spread via the faecal-oral route via contaminated water and/or food and are an important cause of gastroenteritis in young children worldwide. Most astrovirus infections are not severe, self-limited and do not require hospitalization. Disease symptoms can include diarrhoea, followed by nausea, vomiting, fever, malaise and abdominal pain, which last for 3–4 days. The majority of children have acquired astrovirus antibodies by the age of 5 and, looking at the pattern of disease, it suggests that antibodies provide protection through adult life, until the antibody titre begins to decline later in life [90].

Enterovirus Including Poliovirus

Enteroviruses are a genus of positive-stranded RNA viruses in the family *Picornaviridae* with a genome of ~7.5 kb. They are divided in at least 12 species containing over 100 (sero)types. Enteroviruses affect millions of people worldwide each year, are spread through the faecal-oral route, and cause a wide variety of symptoms ranging from mild respiratory illness (common cold), hand, foot and mouth disease, acute hemorrhagic conjunctivitis, aseptic meningitis, myocarditis, severe neonatal sepsis-like disease, and acute flaccid paralysis. Historically, the most prominent member was poliovirus, causing a disabling paralytic illness that has largely been eradicated in most countries through vaccination. Human enterovirus 71 (EV71) epidemics have affected many countries in recent years. Infection commonly causes hand, foot and mouth disease in children, but can result in neurological and cardiorespiratory complications in severe cases. Genotypic changes through inter- and intratypic recombination have been observed, giving rise to serious outbreaks with mortality rate ranging from 10 to 25.7 % [91]. With the emergence of highly pathogenic EV 71 and widespread epidemics, there is great interest in development of an effective EV 71 vaccine and antiviral strategies. In addition, enterovirus 68 has recently emerged as an important cause of severe respiratory disease worldwide [92–96].

As described above, many viruses are able to spread via the faecal-oral route and many more can be detected in human stool in both healthy and diseased adults [36–43, 97–99]. Frequently, the mode of transmission, disease potential and incidence levels of newly recognized viruses detected in stool samples are unknown but potential for food-borne transmission exists. How do we deal with that?

Foodborne Viral Disease Surveillance: Recognition/ Identification

Adequate health crisis management is largely dependent on early detection of potential public health threats. At present, early cluster identification is notoriously difficult as many diseases are not notifiable, diagnostics can be relatively slow and biased for what we know, and clusters are not recognized when patients attend different healthcare facilities. One of the most overlooked but crucial aspects in identifying a potential foodborne related incident is the role that medical practitioners, veterinarians and epidemiologists, in other words the gatekeepers play in recognizing idiopathic disease cases or more than average occurrences of certain disease symptoms [40]. This is not a trivial task as these professionals need to recognize relatively uncommon or completely new infectious diseases, on the basis of changing clinical and epidemiological trends or a "gut-feeling", as syndromic surveillance systems targeting non-respiratory disease are sparse. Integrated networks for syndrome surveillance in combination with routine diagnostic surveillance activities for known pathogens in theory would aid in identification of threats which may otherwise fly under the radar. To date, however, no precise and consistent global baseline syndromic surveillance exists, with the exception of the sentinel surveillance system for influenza. Reliable estimates of the global burden of foodborne viruses are important in order to assess their impact, to advise policy-makers on cost-effective interventions [100], but also to recognize the extraordinary events that signal a potential food-related viral outbreak. The question, however, is how to organise such systems given the ever expanding list of known and potential foodborne viruses.

Classically, many viral pathogens were detected through culture-based and immunological methods, which shifted to molecular detection methods such as polymerase chain reaction (PCR) in more recent years ([101]; Fig. 9.2). The clinical molecular virology field was greatly affected by the development of applications involving viruses that do not proliferate in standard cell cultures and quantitative molecular assays (real time PCRs) that provided medically useful tools in assessing viral load, patient prognosis, treatment response, and antiviral resistance [101]. Currently, the field is moving towards assays that allow detection of multiple viruses. Multiplex PCR assays allow detection of a number of different viruses in a single reaction (e.g. ID-Tag Respiratory Virus Panel Assay identifying influenza A virus [H1, H3, and H5]; influenza B virus; parainfluenza virus types 1, 2, 3, and 4; adenovirus; rhinovirus/enterovirus; RSV A; RSV B; hMPV; and coronavirus [SARS-CoV, NL63, 229E, OC44, and HKU1] [TM Bioscience, Toronto, Canada]). Generic PCR assays are PCR assays specific for a broader taxonomic range than one virus species (e.g. a whole genus or family of viruses), which allows detection of new virus species within already known viral families [41, 102]. These technolo-

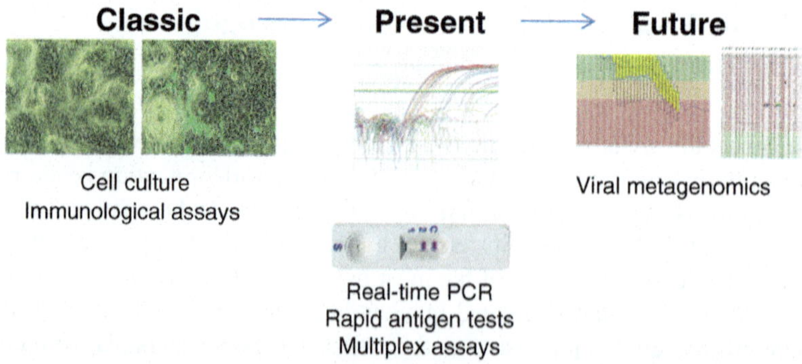

Fig. 9.2 Viral detection methods. Classically, many viral pathogens were detected through culture-based and immunological methods, which shifted to molecular detection methods such as polymerase chain reaction (PCR) in more recent years. Currently, the field is moving towards assays that allow detection of multiple viruses by multiplexing real time PCRs and application of viral metagenomics tools

gies are aiming to decrease time and effort in demonstrating the presence of a known pathogen in clinical samples, although sometimes at the cost of losing some sensitivity [40, 41]. The limitations of the multiplex or generic PCR assays become readily apparent as multiple different viruses or previously unknown viruses can be present in complex biological samples and continuous updating of the assays is required as viruses, especially RNA viruses, are constantly evolving. In addition, in a diagnostic setting discrimination between subtypes or genera of viruses requires additional labour-intensive procedures based on partial genome characterization as is currently done for example for noroviruses, hepatitis A viruses and enteroviruses, for final diagnosis.

With the increasing resolution and use of molecular detection and sequencing, there is great potential for integrated genomic surveillance. The NoroNet network (http://www.rivm.nl/en/Topics/N/NoroNet) in Europe and Asia, and CaliciNet (http://www.cdc.gov/norovirus/reporting/calicinet/index.html) in the US have been developed to aggregate genomic information of noroviruses causing disease outbreaks across the world. In depth bioinformatics analysis of data collected over the course of 10 years has shown the potential merit of genomic surveillance for detection of diffuse foodborne outbreaks [31, 103–105]. Similarly, a regional genomic surveillance database was developed for hepatitis A, enabling cluster analysis as a powerful tool to support outbreak investigations and detect hidden foodborne disease clusters [106]. While these systems target individual pathogens, viral metagenomics tools are a potential candidate to respond to the challenge of obtaining epidemiological estimates on the global disease burden and associated health-related costs of a whole range of (potential) foodborne viruses. Sequence-independent amplification of nucleic acids combined with next-generation sequencing technology and bioinformatics analyses or viral metagenomics is a relatively new promising strategy for rapid identification of pathogens in clinical and public health settings. The detection of viruses using an unselective metagenomics approach has been hampered by the generally small size of virus genomes compared to bacterial or eukaryotic hosts. The detection is facilitated by enriching for viruses using filtration and nuclease treatments to remove bacterial and human nucleic acids whereas viral nucleic acid is retained through protection by the viral capsids and/or membrane envelopes. In contrast to classical molecular detection techniques that identify a single virus species or virus family, viral metagenomics allows the characterization of numerous known pathogens simultaneously and also novel pathogens that elude conventional testing. This approach has already resulted in the identification of a plethora of previously unknown human and animal viruses, many of which have been found in diarrhoea specimens [36–43, 78, 82, 102, 107]. It may approach sensitivity of routine diagnostic real time PCR assays used classically for virus diagnosis [108–110] with the added value of virus type information becoming available simultaneously. In addition, the cost of next-generation sequencing is dropping steadily and steeply each year outpacing Moore's Law. Although computational resources required for analysis of the vast amount of data are often not included in the calculations, the overall costs will likely be able to compete with conventional viral diagnostic molecular methods in the not so distant future in terms of cost and sensitivity, although not yet in speed.

To obtain insight into the baseline circulation of foodborne viruses and the burden of associated disease, a large and systematic set of enteric samples from around the globe from a large range of different individuals with and without (underlying) disease should be analysed. Human exposure to viral infection and susceptibility to virus-associated disease is dependent on numerous factors, including age, lifestyle, diet, geographic location, climate and season, pre-existing immunity and even host microbiome [107]. Furthermore, the human gut virome is not static and will vary over time due to ongoing zoonotic transmission events from animal reservoirs, increasing globalization, changes in food preference, demographic shifts in human populations, and human intervention strategies [25, 40, 41, 107]. However, with the foreseen further implementation of genomic technologies in routine clinical settings, a huge potential surveillance repository is developing. Its validity will depend on the ability to capture meaningful metadata, but the NoroNet and CaliciNet examples have shown that widespread hidden foodborne outbreaks can be detected with sequence data with minimal associated data. Obviously, the validity of such surveillance programs should be carefully evaluated against the current standards to ensure that they provide the necessary information in a timely, efficient, and cost-effective manner [111].

In conjunction with the amount of surveillance data that is required and the huge amount of data generated by next-generation sequencing, the availability of relatively simple user-friendly bioinformatics tools, curated databases of full and partial viral genome sequences, analysis pipelines, and computational infrastructure are crucial and at present largely under development. One example is COMPARE [A COllaborative Management Platform for detection and Analyses of (Re-)emerging and foodborne outbreaks in Europe] which is a collaboration between founding members of the Global Microbial Identifier (GMI) initiative (http://www.globalmicrobialidentifier.org) and institutions with hands-on experience in outbreak detection and response. GMI was established in 2011 with the vision to develop the potential of breakthrough sequencing technologies for the field of infectious diseases through a joint research and development agenda, with applications in clinical and public health laboratories across the world. In order to achieve that long-term goal, the GMI group aims to promote development and deployment of novel applications, data sharing and analysis systems across the diversity of pathogens, health domains and sectors. The COMPARE project is set up to put this vision into action in Europe. It aims to improve rapid identification, containment and mitigation of emerging infectious diseases and foodborne outbreaks by developing a cross-sector and cross-pathogen analytical framework with globally linked data and an information sharing platform that integrates methods for collection, processing and analysing clinical samples with associated (clinical and epidemiological) data with state of the art technologies, such as next generation sequencing, for the generation of actionable information for relevant authorities in human and animal health and food safety.

Assuming the major hurdles towards implementation can be overcome, the combination of sustained virus surveillance (both syndrome and diagnostic) with next generation sequencing approaches and a standardized global analytical framework with associated clinical and epidemiological data would provide insight into (1) pathogens or combinations thereof involved in disease burden, (2) as yet unidentified pathogens and zoonotic events, (3) effects of vaccination or other interventions on incidence levels and whether other pathogens fill the niche that vaccination leaves behind, and (4) geographic difference in virus-associated disease burden. This knowledge would in turn guide development and deployment of vaccines and other intervention strategies. Well, everyone has a wish-list and end-goals … what is the practical translation of viral metagenomics in foodborne viral diseases at present?

Use of Genomics-Based Tools for Foodborne Viral Disease Outbreak Detection: Identification/Characterization

Syndrome surveillance has been used for early detection of disease outbreaks, including food-related incidents, to follow the size, spread, and tempo of outbreaks, to monitor disease trends, and to provide reassurance that an outbreak has not occurred. An example is an outbreak of acute norovirus gastroenteritis in a boarding school in Shanghai in 2012, where a diarrhoea syndrome surveillance system covering a dozen sentinel hospitals in Shanghai reported to the Pudong District Center for Disease Control and Prevention (PDCDC) that more than 100 students at a boarding school had developed symptoms of diarrhoea and vomiting within 3 days [112]. A current practical translation of viral metagenomics, which due to its unselective nature allows the characterization of numerous known pathogens simultaneously, is to use it as an identification tool to unravel the causative viral agent. In the cases in Shanghai, an epidemiological study focusing on a number of viruses (and bacteria) with standard molecular assays subsequently implicated norovirus as the etiological agent [112].

At present, viral metagenomics is mostly used in hindsight to obtain whole viral genome sequences for tracking-and-tracing purposes and to obtain epidemiological information after the virus was identified by more standard molecular assays. In epidemiology, identifying pathways of infectious disease transmission allows amongst others quantification of incubation periods, heterogeneity in transmission rates, and duration of infectiousness, which are important parameters to identify potential points of control and predict future spread of viruses. However, foodborne viral outbreaks are notoriously difficult to recognize, and tracking and tracing potential contacts is logistically challenging and often inconclusive. A variety of data sources can be exploited for attempting to uncover the spatio-temporal dynamics and transmission pathways of a pathogen in a population, by combining disease symptoms, data from contact tracing, results of diagnostic tests and, increasingly, pathogen genetic sequences [113, 114]. Identification of related nucleotide sequences of viruses in patients, also referred to as cluster detection, is an important tool in outbreak investigations in modern day public health and clinical laboratories especially in cases that prove difficult to unravel such as diffuse food-borne outbreaks involving several countries [104]. Norovirus genotype profiles have been used for example to estimate the foodborne proportion of norovirus outbreaks, excluding food handlers as a source of contamination [31, 104, 105]. Preferentially, cluster detection-based approaches and epidemiological inferences are done on whole viral genome sequences, as it provides the most detailed view. Next generation sequencing techniques have been used for tracking purposes in hepatitis A virus (HAV) foodborne outbreaks showing that whole HAV genome analysis offers a more complete genetic characterization of HAV strains than short subgenomic regions [115], although for many viruses partial phylogenetic informative genomic regions can be sufficient for answering the basic tracking-and-tracing questions in an outbreak scenario, with the added advantage of being relatively simple allowing local public health laboratories with limited resources to perform the assays [104, 115].

For informing measures for control of foodborne viral diseases, it is critical to understand the epidemiology in more detail and to accurately identify who-infected-whom, which is usually difficult as data about the location and timing of infections can be incomplete, inaccurate, and compatible with a large number of different transmission scenarios. A number of approaches have been developed that combine genetic and epidemiological data to reconstruct most likely transmission patterns and infection dates [113, 114, 116–119]. These tools may allow for epidemiological studies in real time during outbreaks, which can be used to inform intervention strategies and design control policies [120, 121].

The new developments in data generation with new sequencing possibilities in combination with epidemiological data provide a challenge for existing platforms aiming to enlarge the knowledge on geographical and temporal trends in the emergence and spread of (foodborne) virus infections, such as the ECDC Food- and Waterborne Epidemiology Intelligence Platform (FWD-EPIS) [122], The European Surveillance System managed by ECDC (TESSy; http://ecdc.europa.eu/en/activities/surveillance/Pages/index.aspx), The European Commission Early Warning and

Response System (EWRS) [123] and Rapid Alert System for Food and Feeds (RASFF; http://ec.europa.eu/food/safety/rasff/index_en.htm), NoroNet, and WHO networks among others. In addition to these existing systems, there is a multitude of other existing (inter)national databases and networks that have in common that they are widely accepted and used by the scientific and public health community and authorities for exchange of sequence-based data and other relevant structured and semi-structured information of relevance to human health, animal health and/or food safety. None of these is currently capable of handling the complex data from next generation sequencing platforms, but ensuring interoperability of these databases and compatibility of analytical workflows and data information sharing systems will be crucial in order to ensure translation to actionable data.

Viral Metagenomics and Control of Foodborne Viral Illness: Characterization/Containment

For food safety at present, an integrated system for monitoring of specific food safety threats exists in Europe, which involves sampling and pathogen characterization largely through species specific assays for a subset of major pathogens across the food chain, and linking and analysis of these data to study trends, detect diffuse outbreaks, and monitor effects of control measures [124]. Molecular typing plays a crucial role in this system, but relies among others on the willingness of clinicians to refer patients for laboratory diagnostics and of these laboratories to refer isolates to public health laboratories for typing. The changing clinical practice, with rapid transition from culture-based methods to molecular detection, challenges this decade-old model of disease surveillance [125]. In addition, these surveillance systems are less suited to capture the "new generation" of outbreaks, related with the global food market, as illustrated by recent examples of international diffuse food-borne outbreaks showing the vulnerability of the European population and industry for novel food-borne diseases [25, 106, 126, 127]. The currently used microbiological control criteria are not suitable for monitoring of presence or absence of emerging disease risks, and recent studies have shown vast underestimation of levels of contamination for many human pathogens, but also raise questions about the interpretation of molecular detection data in relation to consumer risk [128, 129].

Improvements in the microbiological safety of food have largely been shaped through response to disease outbreaks. Resources for foodborne diseases have been directed mainly to well-known foodborne pathogens and monitoring in the food chain has been implemented based on a farm-to-fork approach [25] by encouraging improvement of hygiene measures and incorporating Hazard Analysis Critical Control Points (HACCP) principles that identify potential contamination hazards and focus on subsequent control and prevention. The latter requires methods for detection of foodborne pathogens and evidence of their disease association. Most of the microbiological quality control criteria on a global scale rely on standard counts of coliform bacteria as a measure of faecal contamination. Needless to say that these criteria are inadequate for protection against foodborne viruses. Viral metagenomics would theoretically be an option to obtain information regarding viral presence in food. However, microbiological testing of food in general has some limitations as a control option. These are constraints of time, as results are not available until several days after testing as well as difficulties related to sampling as small food samples may not be representative for entire lots, analytical methods and the use of indicator organisms and reference standards. Therefore, it has been argued that there are no practical systems for providing safety or assurance of safety by microbiological end-product testing and viral metagenomics approaches would not change the existing pitfalls.

Concluding Remarks

At present, foodborne pathogen surveillance activities are usually the responsibility of local government departments and are non-existent or at sub-optimal level in both developed and developing countries, are confined to pathogens with known economic impact, and suffer from a lack of integration on a global scale. With the continuing globalization of the food market and changing trends in eating habits [25], it is unsurprising that the general public is faced with an increasing rate of "food safety scares". In order to turn the tide, a huge global effort in virus syndrome and diagnostic surveillance is required, which is justified in the light of global health impact in general, and timely with the development of new metagenomics tools that hold the promise of not only identifying viral pathogens, but possibly the complete microbiome in a single assay. This does not apply to foodborne viral diseases alone. The interrelatedness of animal and human health with global interconnectedness in the twenty-first century is drawing all health related issues together as never before [33]. The combination of sustained pathogen surveillance in animals, humans, plants, environment and food alike with next generation sequencing approaches and a standardized global analytical framework with associated clinical and epidemiological data would provide insight into pathogen incidence, level of co-infections and their correlation to clinical disease instead of focusing on one or a few pathogens as is classically done (Fig. 9.3). This information is crucial in deciding which pathogens provide the most substantial health risk, for evidence-based risk assessments for policy development and to implement preventive measures.

Acknowledgments This work was partially funded by ZonMW TOP project 91213058 and the European Union's Horizon 2020 research and innovation programme under grant agreement No 643476 (COMPARE).

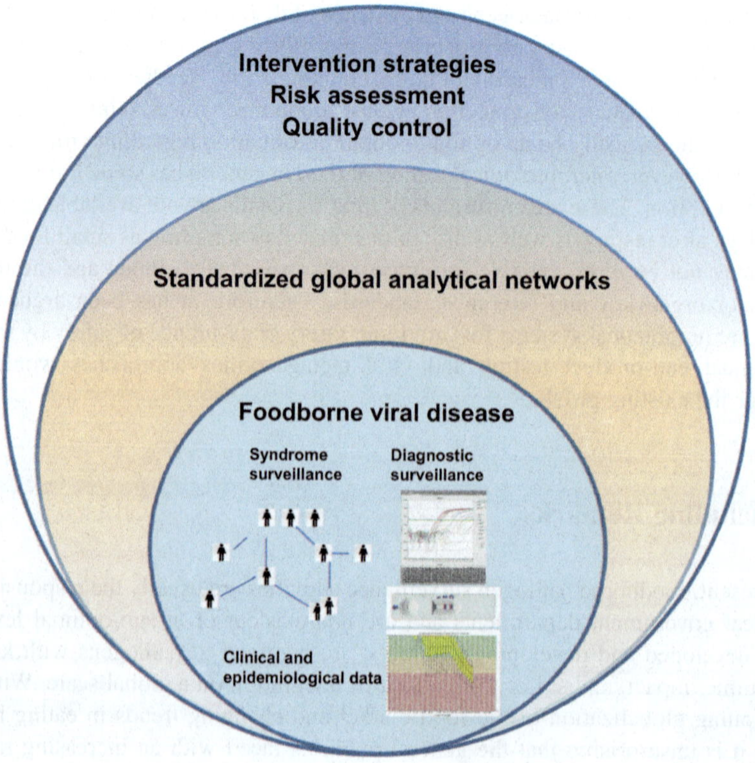

Fig. 9.3 Foodborne viral illness. Schematic overview of the main pillars required for an integrated multidisciplinary approach with a combination of sustained pathogen syndrome and diagnostic surveillance, genomics-based tools, and standardized global analytical networks gathering clinical, epidemiological and genetic data alike would be required to understand the dynamics of foodborne viral infection and to mitigate potential effects of future threats

References

1. Black RE, Cousens S, Johnson HL, Lawn JE, Rudan I, et al. Global, regional, and national causes of child mortality in 2008: a systematic analysis. Lancet. 2010;375:1969–87.
2. Boschi-Pinto C, Velebit L, Shibuya K. Estimating child mortality due to diarrhoea in developing countries. Bull World Health Organ. 2008;86:710–7.
3. Bryce J, Boschi-Pinto C, Shibuya K, Black RE, Group WHOCHER. WHO estimates of the causes of death in children. Lancet. 2005;365:1147–52.
4. Lozano R, Naghavi M, Foreman K, Lim S, Shibuya K, et al. Global and regional mortality from 235 causes of death for 20 age groups in 1990 and 2010: a systematic analysis for the Global Burden of Disease Study 2010. Lancet. 2012;380:2095–128.
5. Murray CJ, Ortblad KF, Guinovart C, Lim SS, Wolock TM, et al. Global, regional, and national incidence and mortality for HIV, tuberculosis, and malaria during 1990–2013: a systematic analysis for the Global Burden of Disease Study 2013. Lancet. 2014;384:1005–70.
6. Parashar UD, Hummelman EG, Bresee JS, Miller MA, Glass RI. Global illness and deaths caused by rotavirus disease in children. Emerg Infect Dis. 2003;9:565–72.

7. Walker CL, Rudan I, Liu L, Nair H, Theodoratou E, et al. Global burden of childhood pneumonia and diarrhoea. Lancet. 2013;381:1405–16.
8. Scallan E, Hoekstra RM, Angulo FJ, Tauxe RV, Widdowson MA, et al. Foodborne illness acquired in the United States—major pathogens. Emerg Infect Dis. 2011;17:7–15.
9. Torgerson PR, de Silva NR, Fevre EM, Kasuga F, Rokni MB, et al. The global burden of foodborne parasitic diseases: an update. Trends Parasitol. 2014;30:20–6.
10. Bishop RF, Davidson GP, Holmes IH, Ruck BJ. Virus particles in epithelial cells of duodenal mucosa from children with acute non-bacterial gastroenteritis. Lancet. 1973;2:1281–3.
11. Flewett TH, Bryden AS, Davies H. Letter: virus diarrhoea in foals and other animals. Vet Rec. 1975;96:JMM.
12. Kapikian AZ, Wyatt RG, Dolin R, Thornhill TS, Kalica AR, et al. Visualization by immune electron microscopy of a 27-nm particle associated with acute infectious nonbacterial gastroenteritis. J Virol. 1972;10:1075–81.
13. Madeley CR, Cosgrove BP. Letter: viruses in infantile gastroenteritis. Lancet. 1975;2:124.
14. Greig JD, Todd EC, Bartleson CA, Michaels BS. Outbreaks where food workers have been implicated in the spread of foodborne disease. Part 1. Description of the problem, methods, and agents involved. J Food Prot. 2007;70:1752–61.
15. Okabayashi T, Yokota S, Ohkoshi Y, Ohuchi H, Yoshida Y, et al. Occurrence of norovirus infections unrelated to norovirus outbreaks in an asymptomatic food handler population. J Clin Microbiol. 2008;46:1985–8.
16. Todd EC, Greig JD, Bartleson CA, Michaels BS. Outbreaks where food workers have been implicated in the spread of foodborne disease. Part 5. Sources of contamination and pathogen excretion from infected persons. J Food Prot. 2008;71:2582–95.
17. Costantini V, Loisy F, Joens L, Le Guyader FS, Saif LJ. Human and animal enteric caliciviruses in oysters from different coastal regions of the United States. Appl Environ Microbiol. 2006;72:1800–9.
18. Iwai M, Hasegawa S, Obara M, Nakamura K, Horimoto E, et al. Continuous presence of noroviruses and sapoviruses in raw sewage reflects infections among inhabitants of Toyama, Japan (2006 to 2008). Appl Environ Microbiol. 2009;75:1264–70.
19. Myrmel M, Berg EM, Grinde B, Rimstad E. Enteric viruses in inlet and outlet samples from sewage treatment plants. J Water Health. 2006;4:197–209.
20. Pommepuy M, Dumas F, Caprais MP, Camus P, Le Mennec C, et al. Sewage impact on shellfish microbial contamination. Water Sci Technol. 2004;50:117–24.
21. Shieh YC, Baric RS, Woods JW, Calci KR. Molecular surveillance of enterovirus and norwalk-like virus in oysters relocated to a municipal-sewage-impacted gulf estuary. Appl Environ Microbiol. 2003;69:7130–6.
22. Victoria M, Guimaraes FR, Fumian TM, Ferreira FF, Vieira CB, et al. One year monitoring of norovirus in a sewage treatment plant in Rio de Janeiro, Brazil. J Water Health. 2010;8:158–65.
23. Wolf S, Hewitt J, Greening GE. Viral multiplex quantitative PCR assays for tracking sources of fecal contamination. Appl Environ Microbiol. 2010;76:1388–94.
24. DuPont HL. Acute infectious diarrhea in immunocompetent adults. N Engl J Med. 2014;370:1532–40.
25. Newell DG, Koopmans M, Verhoef L, Duizer E, Aidara-Kane A, et al. Food-borne diseases—the challenges of 20 years ago still persist while new ones continue to emerge. Int J Food Microbiol. 2010;139 Suppl 1:S3–15.
26. Havelaar AH, Haagsma JA, Mangen MJ, Kemmeren JM, Verhoef LP, et al. Disease burden of foodborne pathogens in the Netherlands, 2009. Int J Food Microbiol. 2012;156:231–8.
27. Scallan E, Griffin PM, Angulo FJ, Tauxe RV, Hoekstra RM. Foodborne illness acquired in the United States—unspecified agents. Emerg Infect Dis. 2011;17:16–22.
28. Ahmed SM, Hall AJ, Robinson AE, Verhoef L, Premkumar P, et al. Global prevalence of norovirus in cases of gastroenteritis: a systematic review and meta-analysis. Lancet Infect Dis. 2014;14:725–30.
29. Mead PS, Slutsker L, Dietz V, McCaig LF, Bresee JS, et al. Food-related illness and death in the United States. Emerg Infect Dis. 1999;5:607–25.

30. Jones KE, Patel NG, Levy MA, Storeygard A, Balk D, et al. Global trends in emerging infectious diseases. Nature. 2008;451:990–3.
31. Verhoef L, Kouyos RD, Vennema H, Kroneman A, Siebenga J, et al. An integrated approach to identifying international foodborne norovirus outbreaks. Emerg Infect Dis. 2011;17:412–8.
32. Fauci AS, Morens DM. The perpetual challenge of infectious diseases. N Engl J Med. 2012;366:454–61.
33. Kuiken T, Leighton FA, Fouchier RA, LeDuc JW, Peiris JS, et al. Public health. Pathogen surveillance in animals. Science. 2005;309:1680–1.
34. Taylor LH, Latham SM, Woolhouse ME. Risk factors for human disease emergence. Philos Trans R Soc Lond B Biol Sci. 2001;356:983–9.
35. Woolhouse ME, Gowtage-Sequeria S. Host range and emerging and reemerging pathogens. Emerg Infect Dis. 2005;11:1842–7.
36. Kapoor A, Li L, Victoria J, Oderinde B, Mason C, et al. Multiple novel astrovirus species in human stool. J Gen Virol. 2009;90:2965–72.
37. Li L, Victoria J, Kapoor A, Blinkova O, Wang C, et al. A novel picornavirus associated with gastroenteritis. J Virol. 2009;83:12002–6.
38. Phan TG, Nordgren J, Ouermi D, Simpore J, Nitiema LW, et al. New astrovirus in human feces from Burkina Faso. J Clin Virol. 2014;60:161–4.
39. Phan TG, Vo NP, Bonkoungou IJ, Kapoor A, Barro N, et al. Acute diarrhea in West African children: diverse enteric viruses and a novel parvovirus genus. J Virol. 2012;86:11024–30.
40. Smits SL, Osterhaus AD. Virus discovery: one step beyond. Curr Opin Virol. 2013;S1879–6257.
41. Osterhaus ADME, Smits SL. Genomics and (re-)emerging viral infections. In: Ginsburg GS, Willard HF, editors. Genomic and personalized medicine. 2nd ed. London: Academic; 2012. p. 1142–54.
42. Smits SL, Rahman M, Schapendonk CM, van Leeuwen M, Faruque AS, et al. Calicivirus from novel recovirus genogroup in human diarrhea, Bangladesh. Emerg Infect Dis. 2012;18:1192–5.
43. Smits SL, Schapendonk CM, van Beek J, Vennema H, Schurch AC, et al. New viruses in idiopathic human diarrhea cases, the Netherlands. Emerg Infect Dis. 2014;20:1218–22.
44. Tate JE, Burton AH, Boschi-Pinto C, Steele AD, Duque J, et al. 2008 estimate of worldwide rotavirus-associated mortality in children younger than 5 years before the introduction of universal rotavirus vaccination programmes: a systematic review and meta-analysis. Lancet Infect Dis. 2012;12:136–41.
45. Desselberger U. Rotaviruses. Virus Res. 2014;190:75–96.
46. Clarke E, Desselberger U. Correlates of protection against human rotavirus disease and the factors influencing protection in low-income settings. Mucosal Immunol. 2015;8:1–17.
47. Ball JM, Tian P, Zeng CQ, Morris AP, Estes MK. Age-dependent diarrhea induced by a rotaviral nonstructural glycoprotein. Science. 1996;272:101–4.
48. Koopmans M, Duizer E. Foodborne viruses: an emerging problem. Int J Food Microbiol. 2004;90:23–41.
49. Ward RL, Bernstein DI, Young EC, Sherwood JR, Knowlton DR, et al. Human rotavirus studies in volunteers: determination of infectious dose and serological response to infection. J Infect Dis. 1986;154:871–80.
50. Atchison C, Iturriza-Gomara M, Tam C, Lopman B. Spatiotemporal dynamics of rotavirus disease in Europe: can climate or demographic variability explain the patterns observed. Pediatr Infect Dis J. 2010;29:566–8.
51. Atchison C, Lopman B, Edmunds WJ. Modelling the seasonality of rotavirus disease and the impact of vaccination in England and Wales. Vaccine. 2010;28:3118–26.
52. Levy K, Hubbard AE, Eisenberg JN. Seasonality of rotavirus disease in the tropics: a systematic review and meta-analysis. Int J Epidemiol. 2009;38:1487–96.
53. Esposito DH, Holman RC, Haberling DL, Tate JE, Podewils LJ, et al. Baseline estimates of diarrhea-associated mortality among United States children before rotavirus vaccine introduction. Pediatr Infect Dis J. 2011;30:942–7.

54. Fischer TK, Nielsen NM, Wohlfahrt J, Paerregaard A. Incidence and cost of rotavirus hospitalizations in Denmark. Emerg Infect Dis. 2007;13:855–9.
55. Fischer TK, Viboud C, Parashar U, Malek M, Steiner C, et al. Hospitalizations and deaths from diarrhea and rotavirus among children <5 years of age in the United States, 1993–2003. J Infect Dis. 2007;195:1117–25.
56. Glass RI, Bresee JS, Parashar U, Miller M, Gentsch JR. Rotavirus vaccines at the threshold. Nat Med. 1997;3:1324–5.
57. Parashar UD, Holman RC, Clarke MJ, Bresee JS, Glass RI. Hospitalizations associated with rotavirus diarrhea in the United States, 1993 through 1995: surveillance based on the new ICD-9-CM rotavirus-specific diagnostic code. J Infect Dis. 1998;177:13–7.
58. Soriano-Gabarro M, Mrukowicz J, Vesikari T, Verstraeten T. Burden of rotavirus disease in European Union countries. Pediatr Infect Dis J. 2006;25:S7–11.
59. Van Damme P, Giaquinto C, Huet F, Gothefors L, Maxwell M, et al. Multicenter prospective study of the burden of rotavirus acute gastroenteritis in Europe, 2004–2005: the REVEAL study. J Infect Dis. 2007;195 Suppl 1:S4–16.
60. Simpson E, Wittet S, Bonilla J, Gamazina K, Cooley L, et al. Use of formative research in developing a knowledge translation approach to rotavirus vaccine introduction in developing countries. BMC Public Health. 2007;7:281.
61. Vinje J. Advances in laboratory methods for detection and typing of norovirus. J Clin Microbiol. 2014;53:373–81.
62. Ramani S, Atmar RL, Estes MK. Epidemiology of human noroviruses and updates on vaccine development. Curr Opin Gastroenterol. 2014;30:25–33.
63. Bresee JS, Marcus R, Venezia RA, Keene WE, Morse D, et al. The etiology of severe acute gastroenteritis among adults visiting emergency departments in the United States. J Infect Dis. 2012;205:1374–81.
64. Gastanaduy PA, Hall AJ, Curns AT, Parashar UD, Lopman BA. Burden of norovirus gastroenteritis in the ambulatory setting—United States, 2001–2009. J Infect Dis. 2013;207:1058–65.
65. Desai R, Hall AJ, Lopman BA, Shimshoni Y, Rennick M, et al. Norovirus disease surveillance using Google Internet query share data. Clin Infect Dis. 2012;55:e75–8.
66. Desai R, Hembree CD, Handel A, Matthews JE, Dickey BW, et al. Severe outcomes are associated with genogroup 2 genotype 4 norovirus outbreaks: a systematic literature review. Clin Infect Dis. 2012;55:189–93.
67. Hall AJ, Lopman BA, Payne DC, Patel MM, Gastanaduy PA, et al. Norovirus disease in the United States. Emerg Infect Dis. 2013;19:1198–205.
68. Bok K, Green KY. Norovirus gastroenteritis in immunocompromised patients. N Engl J Med. 2012;367:2126–32.
69. Koo HL, Neill FH, Estes MK, Munoz FM, Cameron A, et al. Noroviruses: the most common pediatric viral enteric pathogen at a large university hospital after introduction of rotavirus vaccination. J Pediatric Infect Dis Soc. 2013;2:57–60.
70. Patel MM, Widdowson MA, Glass RI, Akazawa K, Vinje J, et al. Systematic literature review of role of noroviruses in sporadic gastroenteritis. Emerg Infect Dis. 2008;14:1224–31.
71. Payne DC, Vinje J, Szilagyi PG, Edwards KM, Staat MA, et al. Norovirus and medically attended gastroenteritis in U.S. children. N Engl J Med. 2013;368:1121–30.
72. Bartsch SM, Lopman BA, Hall AJ, Parashar UD, Lee BY. The potential economic value of a human norovirus vaccine for the United States. Vaccine. 2012;30:7097–104.
73. Jacobsen KH. Hepatitis A virus in West Africa: is an epidemiological transition beginning? Niger Med J. 2014;55:279–84.
74. Jacobsen KH, Koopman JS. The effects of socioeconomic development on worldwide hepatitis A virus seroprevalence patterns. Int J Epidemiol. 2005;34:600–9.
75. Jacobsen KH, Wiersma ST. Hepatitis A virus seroprevalence by age and world region, 1990 and 2005. Vaccine. 2010;28:6653–7.
76. Ciocca M. Clinical course and consequences of hepatitis A infection. Vaccine. 2000;18 Suppl 1:S71–4.

77. Wasley A, Fiore A, Bell BP. Hepatitis A in the era of vaccination. Epidemiol Rev. 2006;28:101–11.
78. Bodewes R, van der Giessen J, Haagmans BL, Osterhaus AD, Smits SL. Identification of multiple novel viruses, including a parvovirus and a hepevirus, in feces of red foxes. J Virol. 2013;87:7758–64.
79. Geng J, Fu H, Wang L, Bu Q, Liu P, et al. Phylogenetic analysis of the full genome of rabbit hepatitis E virus (rbHEV) and molecular biologic study on the possibility of cross species transmission of rbHEV. Infect Genet Evol. 2011;11:2020–5.
80. Johne R, Heckel G, Plenge-Bonig A, Kindler E, Maresch C, et al. Novel hepatitis E virus genotype in Norway rats, Germany. Emerg Infect Dis. 2010;16:1452–5.
81. Johne R, Reetz J, Ulrich RG, Machnowska P, Sachsenroder J, et al. An ORF1-rearranged hepatitis E virus derived from a chronically infected patient efficiently replicates in cell culture. J Viral Hepat. 2014;21:447–56.
82. Raj VS, Smits SL, Pas SD, Provacia LB, Moorman-Roest H, et al. Novel hepatitis E virus in ferrets, the Netherlands. Emerg Infect Dis. 2012;18:1369–70.
83. Takahashi M, Nishizawa T, Sato H, Sato Y, Jirintai, et al. Analysis of the full-length genome of a hepatitis E virus isolate obtained from a wild boar in Japan that is classifiable into a novel genotype. J Gen Virol. 2011;92:902–8.
84. Zhao C, Ma Z, Harrison TJ, Feng R, Zhang C, et al. A novel genotype of hepatitis E virus prevalent among farmed rabbits in China. J Med Virol. 2009;81:1371–9.
85. Meng XJ, Anderson DA, Arankalle VA, Emerson SU, Harrison TJ, et al. Hepeviridae. In: King AMQ, Adams MJ, Carstens EB, Lefkowitz EJ, editors. Virus taxonomy—ninth report of the International Committee on Taxonomy of Viruses. London: Elsevier; 2012. p. 1021–8.
86. Kumar S, Subhadra S, Singh B, Panda BK. Hepatitis E virus: the current scenario. Int J Infect Dis. 2013;17:e228–33.
87. Zhu FC, Zhang J, Zhang XF, Zhou C, Wang ZZ, et al. Efficacy and safety of a recombinant hepatitis E vaccine in healthy adults: a large-scale, randomised, double-blind placebo-controlled, phase 3 trial. Lancet. 2010;376:895–902.
88. Harrach B, Benkö M, Both GW, Brown M, Davison AJ, et al. Adenoviridae. In: King AMQ, Adams MJ, Carstens EB, Lefkowitz EJ, editors. Virus taxonomy—ninth report of the International Committee on Taxonomy of Viruses. London: Elsevier; 2012. p. 108–41.
89. Fratamico PM, Bhunia AK, Smith JL. Foodborne pathogens: microbiology and molecular biology. Chapter: Foodborne and Waterborne Enteric Viruses, Publisher: Caister Academic Press, Editors: P.M. Fratamico, A.K. Bhunia, J.L. Smith, 2005. pp. 121–143.
90. Bosch A, Guix S, Krishna NK, Méndez E, Monroe SS, et al. Astroviridae. In: King AMQ, Adams MJ, Carstens EB, Lefkowitz EJ, editors. Virus taxonomy—ninth report of the International Committee on Taxonomy of Viruses. London: Elsevier; 2012. p. 953–9.
91. Yip CC, Lau SK, Woo PC, Yuen KY. Human enterovirus 71 epidemics: what's next? Emerg Health Threats J. 2013;6:19780.
92. Imamura T, Fuji N, Suzuki A, Tamaki R, Saito M, et al. Enterovirus 68 among children with severe acute respiratory infection, the Philippines. Emerg Infect Dis. 2011;17:1430–5.
93. Jacobson LM, Redd JT, Schneider E, Lu X, Chern SW, et al. Outbreak of lower respiratory tract illness associated with human enterovirus 68 among American Indian children. Pediatr Infect Dis J. 2012;31:309–12.
94. Jaramillo-Gutierrez G, Benschop KS, Claas EC, de Jong AS, van Loon AM, et al. September through October 2010 multi-centre study in the Netherlands examining laboratory ability to detect enterovirus 68, an emerging respiratory pathogen. J Virol Methods. 2013;190:53–62.
95. Kreuter JD, Barnes A, McCarthy JE, Schwartzman JD, Oberste MS, et al. A fatal central nervous system enterovirus 68 infection. Arch Pathol Lab Med. 2011;135:793–6.
96. Meijer A, van der Sanden S, Snijders BE, Jaramillo-Gutierrez G, Bont L, et al. Emergence and epidemic occurrence of enterovirus 68 respiratory infections in The Netherlands in 2010. Virology. 2012;423:49–57.

97. Breitbart M, Hewson I, Felts B, Mahaffy JM, Nulton J, et al. Metagenomic analyses of an uncultured viral community from human feces. J Bacteriol. 2003;185:6220–3.
98. Yu G, Greninger AL, Isa P, Phan TG, Martinez MA, et al. Discovery of a novel polyomavirus in acute diarrheal samples from children. PLoS One. 2012;7, e49449.
99. Zhang T, Breitbart M, Lee WH, Run JQ, Wei CL, et al. RNA viral community in human feces: prevalence of plant pathogenic viruses. PLoS Biol. 2006;4:e3.
100. Kuchenmuller T, Hird S, Stein C, Kramarz P, Nanda A, et al. Estimating the global burden of foodborne diseases—a collaborative effort. Euro Surveill. 2009;14.
101. Leland DS, Ginocchio CC. Role of cell culture for virus detection in the age of technology. Clin Microbiol Rev. 2007;20:49–78.
102. Smits SL, van Leeuwen M, Kuiken T, Hammer AS, Simon JH, et al. Identification and characterization of deer astroviruses. J Gen Virol. 2010;91:2719–22.
103. Verhoef L, Koopmans M, van Pelt W, Duizer E, Haagsma J, et al. The estimated disease burden of norovirus in The Netherlands. Epidemiol Infect. 2013;141:496–506.
104. Verhoef L, Williams KP, Kroneman A, Sobral B, van Pelt W, et al. Selection of a phylogenetically informative region of the norovirus genome for outbreak linkage. Virus Genes. 2012;44:8–18.
105. Verhoef LP, Kroneman A, van Duynhoven Y, Boshuizen H, van Pelt W, et al. Selection tool for foodborne norovirus outbreaks. Emerg Infect Dis. 2009;15:31–8.
106. Petrignani M, Verhoef L, Vennema H, van Hunen R, Baas D, et al. Underdiagnosis of foodborne hepatitis A, The Netherlands, 2008–2010. Emerg Infect Dis. 2014;20:596–602.
107. Delwart E. A roadmap to the human virome. PLoS Pathog. 2013;9, e1003146.
108. de Vries M, Oude Munnink BB, Deijs M, Canuti M, Koekkoek SM, et al. Performance of VIDISCA-454 in feces-suspensions and serum. Viruses. 2012;4:1328–34.
109. Prachayangprecha S, Schapendonk CM, Koopmans MP, Osterhaus AD, Schurch AC, et al. Exploring the potential of next-generation sequencing in detection of respiratory viruses. J Clin Microbiol. 2014;52:3722–30.
110. Yang J, Yang F, Ren L, Xiong Z, Wu Z, et al. Unbiased parallel detection of viral pathogens in clinical samples by use of a metagenomic approach. J Clin Microbiol. 2011;49:3463–9.
111. Drewe JA, Hoinville LJ, Cook AJ, Floyd T, Stark KD. Evaluation of animal and public health surveillance systems: a systematic review. Epidemiol Infect. 2012;140:575–90.
112. Xue C, Fu Y, Zhu W, Fei Y, Zhu L, et al. An outbreak of acute norovirus gastroenteritis in a boarding school in Shanghai: a retrospective cohort study. BMC Public Health. 2014;14:1092.
113. Jombart T, Aanensen DM, Baguelin M, Birrell P, Cauchemez S, et al. OutbreakTools: a new platform for disease outbreak analysis using the R software. Epidemics. 2014;7:28–34.
114. Jombart T, Cori A, Didelot X, Cauchemez S, Fraser C, et al. Bayesian reconstruction of disease outbreaks by combining epidemiologic and genomic data. PLoS Comput Biol. 2014;10, e1003457.
115. Vaughan G, Xia G, Forbi JC, Purdy MA, Rossi LM, et al. Genetic relatedness among hepatitis A virus strains associated with food-borne outbreaks. PLoS One. 2013;8, e74546.
116. Mollentze N, Nel LH, Townsend S, le Roux K, Hampson K, et al. A Bayesian approach for inferring the dynamics of partially observed endemic infectious diseases from space-time-genetic data. Proc Biol Sci. 2014;281:20133251.
117. Morelli MJ, Thebaud G, Chadoeuf J, King DP, Haydon DT, et al. A Bayesian inference framework to reconstruct transmission trees using epidemiological and genetic data. PLoS Comput Biol. 2012;8, e1002768.
118. Teunis P, Heijne JC, Sukhrie F, van Eijkeren J, Koopmans M, et al. Infectious disease transmission as a forensic problem: who infected whom? J R Soc Interface. 2013;10:20120955.
119. Ypma RJ, Bataille AM, Stegeman A, Koch G, Wallinga J, et al. Unravelling transmission trees of infectious diseases by combining genetic and epidemiological data. Proc Biol Sci. 2012;279:444–50.
120. Ferguson NM, Donnelly CA, Anderson RM. Transmission intensity and impact of control policies on the foot and mouth epidemic in Great Britain. Nature. 2001;413:542–8.

121. Keeling MJ, Woolhouse ME, May RM, Davies G, Grenfell BT. Modelling vaccination strategies against foot-and-mouth disease. Nature. 2003;421:136–42.
122. Gossner C. ECDC launches the second version of the EPIS-FWD platform. Euro Surveill. 2013;18.
123. Guglielmetti P, Coulombier D, Thinus G, Van Loock F, Schreck S. The early warning and response system for communicable diseases in the EU: an overview from 1999 to 2005. Euro Surveill. 2006;11:215–20.
124. EFSA and ECDC. Scientific report of EFSA and ECDC—the European Union summary report on trends and sources of zoonoses, zoonotic agents and food-borne outbreaks in 2012. EFSA J. 2014;12:3547.
125. Jones TF, Gerner-Smidt P. Nonculture diagnostic tests for enteric diseases. Emerg Infect Dis. 2012;18:513–4.
126. Aboutaleb N, Kuijper EJ, van Dissel JT. Emerging infectious colitis. Curr Opin Gastroenterol. 2014;30:106–15.
127. Grad YH, Lipsitch M, Feldgarden M, Arachchi HM, Cerqueira GC, et al. Genomic epidemiology of the *Escherichia coli* O104:H4 outbreaks in Europe, 2011. Proc Natl Acad Sci U S A. 2012;109:3065–70.
128. Stals A, Baert L, De Keuckelaere A, Van Coillie E, Uyttendaele M. Evaluation of a norovirus detection methodology for ready-to-eat foods. Int J Food Microbiol. 2011;145:420–5.
129. Stals A, Baert L, Van Coillie E, Uyttendaele M. Evaluation of a norovirus detection methodology for soft red fruits. Food Microbiol. 2011;28:52–8.

Chapter 10
Transcriptomics and Proteomics of Foodborne Bacterial Pathogens

Joelle K. Salazar, Yun Wang, and Wei Zhang

Introduction

This chapter focuses on the applied transcriptomics and proteomics of foodborne bacterial pathogens, including *Salmonella enterica*, *Listeria monocytogenes*, and *Escherichia coli*. The first part of the chapter discusses the transcriptomic techniques of RNA-seq and chromo immunoprecipitation (ChIP)-seq as related to foodborne pathogen studies. The second section of the chapter describes the proteomic techniques mass spectrometry and protein microarrays. Special attention is given to methodology of each technology, applications of the techniques from the recent literature, and specific limitations and challenges that each technique faces, along with future perspectives.

Transcriptomics

RNA Sequencing

Regulation of gene expression in bacteria is paramount for bacteria to adapt to environmental stresses, and allow persistence, pathogenicity, and virulence in a host. DNA microarray technology has been a popular tool used by researchers to study

J.K. Salazar (✉) • Y. Wang
Division of Food Processing Science and Technology, U. S. Food and Drug Administration, Bedford Park, IL 60501, USA
e-mail: joelle.salazar@fda.hhs.gov; yun.wang2@fda.hhs.gov

W. Zhang
Department of Food Science and Nutrition, Illinois Institute of Technology, Bedford Park, IL 60501, USA
e-mail: zhangw@iit.edu

X. Deng et al. (eds.), *Applied Genomics of Foodborne Pathogens*, Food Microbiology and Food Safety, DOI 10.1007/978-3-319-43751-4_10

the transcriptome and the regulation of gene expression in microorganisms. A microarray consists of millions of nucleotide probes attached to the surface of a small glass slide. This technology can be used to comparatively analyze the transcriptomes of two different organisms, or the same organism under different experimental conditions or treatments. In recent years, this technology is no longer the most desirable in the toolbox for bacterial transcriptomic characterization due to many inherent technical limitations, such as the need of a reference genome for probe design and the high cost and complexity for manufacturing custom microarrays that fit individual research needs. A newer technology, referred to as RNA sequencing or 'RNA-seq', is built upon next-generation sequencing (NGS) platforms, which was first introduced in 2008 to study the transcriptomes of *Saccharomyces cerevisiae* [1] and *Schizosaccharomyces pombe* [2]. RNA-seq requires simple steps of total RNA extraction of the organism of interest, cDNA synthesis, and preparation of the cDNA depending on the desired sequencing platform. The first application of RNA-seq on a prokaryotic organism was for the discovery of new genes in *Sinorhizobium meliloti* [3]. Since then, numerous studies have deployed RNA-seq technology to characterize transcriptional landscapes of bacteria, including foodborne pathogens such as *Salmonella enterica*, *Listeria monocytogenes*, *Escherichia coli*, *Campylobacter jejuni*, and others. RNA-seq has aided in uncovering transcriptional regulations and thus a myriad of information about the persistence, pathogenicity, and virulence of these foodborne bacterial pathogens.

Methodology of RNA-Seq Technology

RNA-seq technology for bacteria starts with extraction of total RNA from a culture, either by using a commercially available kit, or by standard phenol/guanidine isothiocyanate methods. Contaminating DNA is removed by digestion with DNAse to obtain purified RNA. Since rRNA and tRNA constitute roughly 95 % of total bacterial RNA, these functional RNAs should be removed before sequencing using a commercially available kit or other published methods [4–6]. Next, if transcript directionality is not a concern in a particular study, the mRNA is converted to double stranded cDNA by reverse transcriptase. In the case where directionality is of a concern, only first strand cDNA synthesis is conducted [7, 8]. The cDNA is then prepared according to the specific sequencing platform to be used. For example, in Illumina sequencing, cDNA fragments are end-repaired, followed by adenylation of 3′ ends. Specific adapter sequences are ligated to the 3′ ends and PCR is used to amplify the fragments using primers with complementary sequences to the adapters. For multiplexing, specific index sequences are added to the primer sequences; indices allow for differentiation of samples when multiple samples are analyzed together. Figure 10.1 depicts an overview of a typical RNA-seq sample preparation.

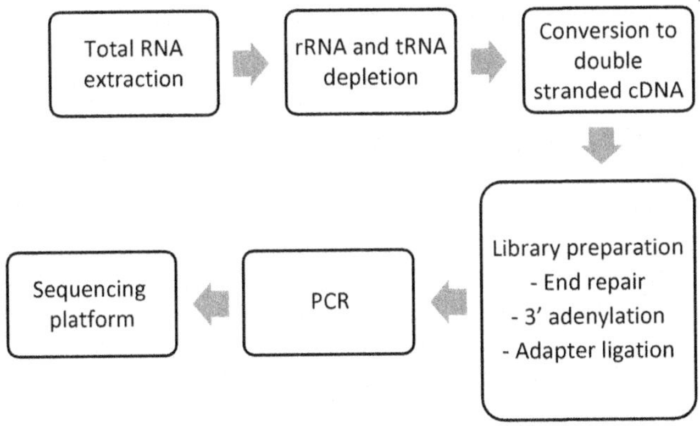

Fig. 10.1 Overview of RNA-seq sample preparation

Sequencing Platforms

As of today, the three most commonly used NGS platforms include Roche 454, Applied Biosciences SOLiD, and Illumina, although Illumina by far has the monopoly. Roche 454 technology employs emulsion PCR technology coupled with pyrosequencing. Applied Biosciences SOLiD sequencing also utilizes emulsion PCR, but uses arrays of beads called microreactors [9]. Illumina sequencing technology, unlike Roche 454 and SOLiD, utilizes isothermal bridge amplification along with fluorescent reversible terminator sequencing in a reaction chamber, or flow cell [10]. DNA molecules are denatured, resulting in single-stranded fragments, which are added to the flow cell. The DNA strands adhere to the flow cell surface via complementary oligonucleotides to one of the adapter sequences. The DNA strands are then clonally amplified by bridge amplification, where each strand folds over and binds to the second complementary oligonucleotide on the flow cell. Clusters are formed by amplification of the DNA molecules. The reverse strands are cleaved and washed off, resulting in only the forward strand remaining on the flow cell. Sequence analysis occurs by the incorporation of fluorescent nucleotides, which bind to their complementary counterpart on the growing strand. For paired-end sequencing, DNA fragments are bridge amplified and the forward strand is cleaved and washed off, followed by sequence analysis as before. The Illumina platform also utilizes multiplexing technology, where multiple samples with different indices can be run together on one flow cell. The Illumina MiSeq system produces four million reads with a read length of 300 bp (Table 10.1). For the Illumina HiSeq 2500 instrument, three billion reads are produced with 150 bp read length.

Analysis of RNA-seq data involves processing raw sequenced reads, mapping the reads to a reference genome, and reporting gene expression. All processed reads include a base call and a quality score. Based on the quality scores and the sequencing platform used, the reads are trimmed to eliminate low-quality reads. Read pro-

Table 10.1 Comparison of NGS platforms

Platform	Roche 454[a]	SOLiD[b]	Illumina[c,d]
Amplification	Emulsion PCR	Emulsion PCR	Bridge amplification
Sequencing chemistry	Pyrosequencing	Ligation	Reversible terminator
Detection method	Chemiluminescence	Fluorescence	Fluorescence
Current read length	800 bp	35 bp	300 bp[c], 150 bp[d]
Reads per run	1 million	1.4 billion	4 million[c], 3 billion[d]
Data generated per run	~1 Gb	~150 Gb	~2 Gb[c], ~600 Gb[d]
Run time	~24 h	8 days	~24 h[c], 11 days[d]
Error rate	0.10 %	~2–4 %	0.80 %[c], 0.26 %[d]

[a]Roche 454 Genome Sequencer FLX
[b]Applied Biosystems SOLiD 5500xl
[c]Illumina MiSeq
[d]Illumina HiSeq 2500

cessing also involves de-multiplexing if different index sequences were used. All processed reads are then mapped to a reference genome, or if no reference genome is available, assembled de novo. Once assembled, gene expression values can be reported; such metrics include RPKM (reads per kilobase per million mapped reads) [11, 12] and gene expression index (GEI) (normalized reads per kilobase) [13]. A number of commercially and publically available software tools are available for RNA-seq analysis. A comprehensive listing of software tools can be found at the Sanger Institute (www.sanger.ac.uk) and Bioconductor (www.bioconductor.org) websites. Some of the main tools for analysis are listed in Table 10.2.

Applications of RNA-Seq

Myriad studies have looked at the transcriptomic profiles of bacteria, including foodborne pathogens, under different treatment conditions (i.e. under stressed conditions or in different food matrices) using DNA microarray technology [32–39]. In recent years, RNA-seq has been the method of choice for studying bacterial global gene expression profiles. For example, Oliver et al. [40] utilized Illumina technology to characterize the transcriptome of *L. monocytogenes* and a mutant strain which lacked sigma factor B, an important transcriptional regulator. The authors discovered that during stationary phase, 96 genes were upregulated in the wild-type strain, showing that these genes are dependent on regulation by sigma factor B.

RNA-seq has been used to study pathogen response to the food matrix environment, providing insights into how these bacteria are able to survive and persist under suboptimal conditions. For instance, Deng et al. [41] studied the transcriptome of *S. enterica* serovar Enteritidis under desiccation and starvation stress in peanut oil. In this study, peanut oil was used as a substitute for peanut butter, a vector linked to multiple outbreaks of salmonellosis. The authors determined that less than 5 % of the Enteritidis genome was transcribed under desiccation stress in peanut oil, as compared with 78 % in Luria-Bertani broth. Some of the genes discov-

Table 10.2 Brief list of bioinformatics tools for analysis of RNA-seq data

Application	Software tool	References
Quality control and read filtering	FastQC	[14]
	FreClu	[15]
	RNA-SeQC	[16]
	RSeQC	[17]
	ShortRead	[18]
	Trimmomatic	[19]
Pre-processing data	DeconRNASeq	
	FLASH	[20]
Short alignment	Bowtie	[21]
	BWA	[22]
	GNUMAP	[23]
	RazerS	[24]
Quantitative analysis	Cufflinks	[25]
	DESeq	[26]
	EGDE-pro	[27]
	DEGseq	[28]
Visualization tools	Artemis	[29]
	IGV	[30]
	GenomeView	[31]

ered to be transcribed under dessication stress encoded proteins for stress response to temperature shifts, including heat shock sigma factor RpoH [42] and an extreme heat and cell envelope stress sigma factor RpoE [43]. Other studies of *Salmonella* in low water activity foods have investigated the transcriptomic response of the bacterium to desiccation stresses. These stress-induced proteins aid in resistance and the survival of the pathogen in low water activity food products. This was the first study of RNA-seq to characterize a bacterial transcriptome associated with a food matrix. In another study, Brankatschk et al. [44] used RNA-seq to study the transcriptome of *S. enterica* serovar Weltevreden during alfalfa sprout colonization. The authors deduced that approximately 4 % of genes were transcribed at higher levels during colonization, including those for attachment, motility, and biofilm formation.

The ability of foodborne pathogenic organisms to resist disinfectants or antimicrobial compounds is also an important mechanism of persistence in the food processing environment. Often, industry settings use disinfectants, sanitizers, and antimicrobial agents to prevent pathogen contamination of food products. Feng et al. [45] used RNA-seq to study the transcriptome of *Cronobacter sakazakii*, a foodborne pathogen linked to outbreaks of infant formula [46], under unfavorable stressed conditions. The study focused on the use of two garlic-derived organosulfur compounds as antimicrobial agents against *C. sakazakii*. The agents were effective in the inactivation of the pathogen; in the presence of the compounds, a set of 133 genes were significantly downregulated. These genes encode proteins with roles in motility, oxidioreductase activity, and biosynthetic processes. Another study by Fox et al. [47] investigated the transcriptional differences between a persistent and a non-persistent strain of *L. monocytogenes* in the presence of a sublethal quantity of a quaternary ammonium compound, benzethonium chloride. Comparison of the gene expression profile of the persistent strain of *L. monocytogenes* in the presence of benzethonium chloride and that of the non-persistent strain revealed that 63 genes were significantly upregulated. These included genes encoding metabolism regulators, transport and binding proteins, cofactor biosynthesis proteins, and osmotic stress proteins. Another study on the transcriptome of a persistent strain of *L. monocytogenes* exposed to biocide stress was conducted by Casey et al. [48]. The study found that the pathogen responded to biocide stress by upregulating genes involved in processes such as peptidoglycan biosynthesis, chemotaxis and motility, and carbohydrate uptake. These studies aid in elucidating the mechanisms by which some foodborne pathogens are capable of surviving in food processing environments in the presence of disinfectants and antimicrobial agents.

RNA-seq has also been a useful tool to study virulence and pathogenicity of foodborne pathogens during infection and colonization within the host. Studies with transcriptome sequencing have been beneficial in discovering the molecular determinants that are required for infection. Taveirne et al. [49] studied the transcriptome of *C. jejuni* during colonization of host chicken cecum. The authors found that over 250 genes were differentially expressed during passage through the host, including genes encoding proteins in biosynthetic processes, energy and metabolism, motility, stress response, and transport. The findings aid in elucidating how this pathogen is capable of surviving and adapting to the host environment. Jin et al. [50] used RNA-seq to study the transcriptome of *E. coli* during infection of bovine mammary epithelial cells. The study concluded that the host cells played a role in immunity and development when challenged with the pathogenic bacteria. Kröger et al. [51] studied the transcriptomic landscape of *S. enterica* Typhimurium under 22 different infection-related conditions, including conditions representing the intracellular life of the pathogen. During the change from normal to stressed conditions, such as osmotic, anaerobic, nitric oxide, and peroxide shocks, the expression of 15–25 % of all genes changed within 10 min. This represents the ability of *S.* Typhimurium to quickly respond to a changing environment. Mraheil et al. [52] studied *L. monocytogenes* during growth in macrophage cells and deduced that 29 regulatory RNAs, including some small non-coding antisense RNAs, were expressed during intracellular infection of the pathogen. The expression of these regulatory RNAs was determined to be necessary for efficient intracellular growth, and aid in the pathogen's ability to switch from an extracellular to intracellular lifestyle.

Limitations and Challenges

Some of the challenges associated with RNA-seq lie in library preparation, efficient mRNA enrichment and issues with polarity in cDNA synthesis, as well as in bioinformatics analysis. Bacterial total RNA consists mainly of rRNA and tRNA, meaning that a large proportion of total RNAs must be removed for efficient transcriptome analysis [53]. mRNAs comprise only 1–5 % of total RNA, so enrichment of mRNA is a challenging and important step. Methods to deplete rRNA and tRNA have been developed, including use of a terminator 5′-phosphate-dependent exonuclease treatment [54, 55], or the use of biotinylated probes that selectively bind rRNA molecules (i.e. Epicenter Ribo-Zero rRNA Removal Kit). Strand-specificity is also a limitation with some cDNA synthesis methods. Generally, enriched single-stranded mRNA molecules are converted to cDNA using random hexamer primers. This does not provide directionality to the resulting double-stranded cDNA molecules. Lack of this information makes some downstream computational analysis difficult (i.e. alignment to a reference genome). Efforts have been made in providing directionality in RNA-seq studies [8, 56, 57]. Another challenge in RNA-seq is how to store, retrieve, and process the large amounts of data resulting from sequencing if sufficient computation power is not available. Once reads are adequately quality checked according to various metrics and trimmed, they must be mapped and aligned to a reference genome, and quantitatively analyzed for differential expression. Different tools are available for each step in the analysis pipeline and results may vary depending on the application used; Williams et al. discusses a thorough analysis of challenges associated with RNA-seq data [58] and presents recommendations to improve sample quality, read alignment, and assigning reads to genes or transcripts.

ChIP Sequencing

Binding of proteins to DNA molecules within a cell can serve many functions: DNA synthesis, regulation of transcription and translation, initiation and inhibition of metabolic pathways, and so on. For foodborne pathogens, protein-DNA interactions are critical for environmental and in-host survival and subsequent virulence. Chromatin immunoprecipitation (ChIP) uses specific antibodies to a protein of interest *in vivo* to precipitate DNA binding partners. ChIP can be coupled to either microarray technology (ChIP-Chip) or NGS technology (ChIP-seq). Some of the first experiments using ChIP technology on bacterial species were ChIP-chip studies, where probes are spotted onto a microarray, and each part of the genome being studied is represented by at least one probe. This is referred to as genome "tiling", which is uniquely suited to the small size of bacterial genomes. Using ChIP-chip, studies have looked at the function of various bacterial proteins such as RacA (a cell segregation protein) in *B. subtilis* [59, 60], H-NS and StpA (nucleoid-associated proteins) in *S.* Typhimurium [61, 62], and H-NS and RNA polymerase in *E. coli* [63–68]. Studies have also uncovered DNA binding partners and consensus sequences of transcription factors such as Fur in *Helicobacter pylori* [69], CodY in *B. subtilis* [70], LexA in *E. coli* [71], and SsrB and HilA in *S.* Typhimurium [72, 73]. Only in the very recent years has ChIP-seq technology been applied to bacteria, including some foodborne pathogens (see Applications).

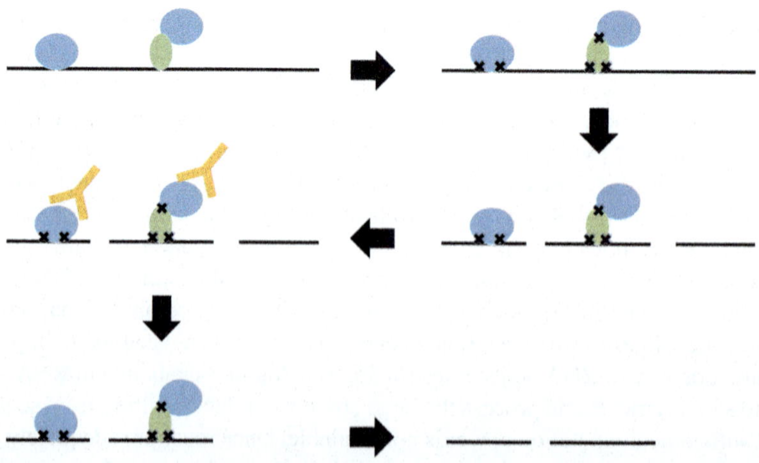

Fig. 10.2 Overview of a ChIP experiment. Protein-DNA complexes are crosslinked (*black* "x") with formaldehyde in vivo. The DNA is sheared by sonication, followed by the addition of an antibody (*yellow*) to the protein of interest (*blue*). The protein-DNA complexes are precipitated and the DNA is purified

Methodology of ChIP-Seq Technology

A typical bacterial ChIP-seq experiment begins with crosslinking protein-DNA interactions *in vivo*. Crosslinking can be achieved with the addition of formaldehyde, a small molecule which readily diffuses into the cell. Proteins are crosslinked directly to DNA, or indirectly to DNA through protein–protein interactions. The cells are then lysed and DNA is fragmented to approximately 200–500 bp through the use of sonication or chemical means. Once purified cell lysates are obtained, an antibody is used to specifically bind to the protein of interest, either directly to the protein or to a tag on the protein such as FLAG-tag [74] or His-tag [75]. The antibody bound DNA-protein complex is then precipitated using Protein A or G beads coupled to centrifugation or filtration techniques. To decrosslink the protein-DNA interactions, the samples are heated at 65 °C overnight or boiled for 10 min. The resulting DNA is purified and analyzed for quality and quantity, generally via qPCR and the Agilent Bioanalyzer instrument (see Fig. 10.2). To prepare DNA samples for sequencing, adapters are ligated to DNA molecules and then PCR is used to amplify the DNA using primers complementary to the adapter sequences. Since a ChIP experiment generally only produces 1–10 ng of DNA (using approximately 10^7 cells), the DNA must be amplified to attain enough material before sequencing. If multiplexing will be used, each sample must be properly labelled using index sequences. A few commercial ChIP sequencing preparation kits are available, including the TruSeq ChIP Sample Preparation Kit (Illumina, Inc.), the NEXTflex ChIP-seq Kit (Bioo Scientific Corp.), and the DNA SMART ChIP-seq Kit (Clontech, Takara Bio Co.) for use with the Illumina NGS platform.

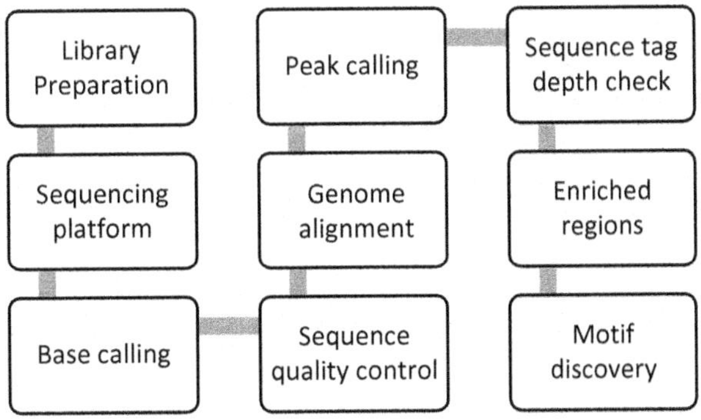

Fig. 10.3 Basic steps in a ChIP-seq experiment

Initial analysis of ChIP-seq data is not unlike that of RNA-seq: the raw sequenced reads are processed and checked for quality, and then the reads are mapped to a reference genome (Fig. 10.3). Since ChIP involves the enrichment of DNA regions that were directly or indirectly bound to the protein of interest, "peaks" or regions of the genome with more coverage as compared to a control sample are identified. The most widely used controls for a ChIP-seq experiment are input (DNA taken prior to immunoprecipitation) and mock (treated the same way as the immunoprecipitated sample but without antibody) samples. The most common software tool to identify peaks is Model-based Analysis for ChIP-seq or MACS [76]. MACS uses a dynamic Poisson distribution to identify peaks and the software can be used with or without a control sample. Other peak identification software include PeakSeq [77] which takes into account the differences in the mappability (i.e. alignment of reads, of the genome), FindPeaks [78], and CisGenome [79]. After peaks are discovered, they must be annotated to determine DNA target regions of the protein of interest. These targets can be genes, up-stream or down-stream regions of genes, promoter sequences, or regions inside genes [80, 81]. If the recognition sequence of the protein of interest is known, this can aid in verifying the sequencing data. Data can also be verified by electrophoresis mobility shift assays [82] and DNA footprint assays [83].

Applications of ChIP-Seq

Although there is no technical hurdle in applying ChIP-seq to prokaryotes, studies are mainly focused on eukaryotic organisms. Only a handful of ChIP-seq studies have been published on bacteria, and even less on foodborne pathogens. Those that have been published focused mainly on proteins and transcriptional regulators involved in environmental survival, and virulence and pathogenicity within a host. Responding to environmental stimuli and maintaining cell homeostasis are critical for pathogen survival in various niches. Davies et al. [84] used ChIP-seq to study the ferric uptake regulator, Fur, in *Vibrio cholera*. Fur regulates iron transport for enzymatic functions within the cell and is a key factor in the maintenance of homeostasis. The authors discovered many novel Fur-regulated genes, including those involved in multidrug resistance, chemotaxis, and transport. Since Fur does not bind to canonical sequences, ChIP-seq was able to predict DNA binding sites based on peaks associated with open reading frames and sRNAs. [85] studied AraC, a transcriptional activator of genes involved in arabinose metabolism in *E. coli* and *S. enterica* and other *Enterobacteriaceae*. Using ChIP-seq with *S.* Typhimurium, the study uncovered two novel AraC-activated genes, *araT* and *araU*, which are likely involved in the transport and metabolism of arabinosides, suggesting that *S. enterica* can use arabinosides as a carbon source. The authors compared the DNA binding sites of AraC in *S. enterica* and *E. coli* with those of other *Enterobacteriaceae*, including *Enterobacter* spp., *Klebsiella pneumonia*, *Citrobacter rodentium*, and *Cronobacter sakazakii*, and identified a conserved consensus sequence. Fitzgerald et al. [86] published a ChIP-seq study looking at flagella regulation in *E. coli* K12. Although K12 is nonpathogenic, FlhDC is conserved as the master flagellar regulator in pathogenic *E. coli*, *S. enterica*, and closely related enteric bacteria. Flagella are known to play a role in bacterial chemotaxis, attachment to surfaces, and in host cell colonization. Two regulators involved in flagellar synthesis, FlhDC and FliA, were studied. It has been suggested that these regulators may also be involved in regulating non-flagellar genes. ChIP-seq revealed four novel FlhDC binding sites, three in intergenic regions and one inside the gene *csgC*, and 52 novel FliA binding sites, 30 of which are inside genes.

Three studies published in 2013 used ChIP-seq with *S. enterica* to investigate the regulation of *Salmonella* pathogenicity islands (SPI), survival and colonization in host cells, and host immune evasion. Wang et al. [87] used ChIP-seq to study Fis, an important nucleoid-associated protein which functions as a regulator of transcription and virulence in *S. enterica*. The authors identified 1646 Fis-regulated genes in *S.* Typhimurium, including 63 (of a total of 94) SPI-1 and SPI-2 genes. Regulated genes included nine genes encoding effector proteins which are transported by the type III secretion system into the host cell, 37 genes encoding needle complex proteins, and six genes encoding regulators. The results were further verified by electrophoresis mobility shift assays and macrophage and epithelial infection assays. A *S.* Typhimurium Δ*fis* mutant exhibited a significant decrease in the ability to invade these two cell types, demonstrating that Fis is necessary for invasion and intracellular replication. Another study looking at SPI-1 regulation in *S.* Typhimurium was conducted by Petrone et al. [88]. ChIP-seq was used to study the transcriptional activator HilD, which is known to regulate SPI-1 genes, under conditions associated with an intracellular host environment. The authors uncovered 17 HilD-binding regions, of which 11 were novel and nine are located outside of SPI-1. *lpxR*, one of the novel regulated genes, encodes a protein responsible for removing the 3′ acyloxyacyl group of lipid A on LPS [89]. This allows *S.* Typhimurium to evade the innate immune response, allowing the pathogen to survive within macrophage cells [90, 91]. A ChIP-seq study on *S.* Typhimurium by Perkins et al. [92] determined that OmpR plays a role in host cell colonization. OmpR has previously been determined to play roles in the regulation of transcription in response to environmental stimuli such as osmotic stress. ChIP-seq revealed 43 intergenic peaks and seven novel

OmpR-regulated genes encoding proteins which play a role in sialic acid transport, an oxidioreductase, and an N-acetylmannosamine-6-P epimerase. These proteins may aid in the scavenging of nutrients during host cell colonization by *S. enterica.*

Limitations and Challenges

Some of the challenges associated with ChIP-seq include antibody quality, the amount of material required for sequencing, adequate controls, and data analysis. One of the most important aspects when designing a ChIP-seq experiment is to ensure that the antibody used will precipitate the protein of interest *in vivo*. Although there are various commercial antibodies available, not all are validated for ChIP-seq studies. All antibodies should be verified for effectiveness before beginning a ChIP-seq experiment, which is time consuming. If the protein of interest does not have an associated antibody, it may be necessary to add a tag and precipitate with a commercially available monoclonal antibody. When tags are added, changes in the conformation and function of the protein may occur, possibly leading to the insufficient binding of DNA and protein partners. Another drawback with ChIP-seq includes the amount of material required. Generally, 10–50 ng of sample is required for a sequencing run. Although amplification via PCR is seen in most protocols (with 15–18 cycles), this amplification can create bias and cycles should be limited. Therefore, it may be necessary to pool various ChIP-seq studies together to acquire enough material. To ensure that a ChIP-seq experiment is working, adequate controls should be used, especially in the case of a protein where no target DNA partners are known. A control should consist of a protein or transcription factor in which some DNA binding partners are known and enrichment can be verified using qPCR. Another limitation of ChIP-seq is the storage, manipulation, and analysis of the massive amount of data generated from a sequencing run if adequate computational power is not available. There is also no community consensus on what data should be saved (i.e. raw image data, raw sequence reads) or how precisely data should be analyzed. For the non-bioinformatician, data analysis can always be challenging as many current software programs are not user-friendly.

Conclusions and Future Perspectives

RNA-seq technology, in only a short period of time, has revolutionized how we study bacterial transcriptomics. This technology has been beneficial in the annotation of genomes and the identification of new genes. Pathogen responses to environmental stimuli and stresses, along with the switch from an extracellular to intracellular lifestyle and subsequent pathogenicity have all been analyzed. Results from these studies have identified novel virulence factors, and have shed light on how regulation occurs in a variety of foodborne pathogens. Undoubtedly, RNA-seq experimental design and bioinformatics analysis will continue to improve in the coming years. Analysis of the data obtained by RNA-seq is now possible by researchers with only modest bioinformatics experiences. Freely available software tools and interfaces, such as Artemis [93] and Galaxy [94], have made the analytical process simpler and more intuitive. New ways to utilize RNA-seq technology will most likely be developed. Such methods will include transcriptomic studies of complex bacterial populations and communities. In addition, the first few manuscripts studying dual RNA-seq technology [95], in which the transcriptomes of both pathogen and host are assayed simultaneously, have just been published; this technology will aid in the understanding of host-pathogen relationships, colonization, and pathogenicity.

ChIP-seq provides unprecedented insight into the transcriptional landscape and cascades of a microorganism. Although few studies to date have used ChIP-seq to specifically study foodborne pathogens, more studies will undoubtedly be published in the future. NGS technology is becoming more affordable for investigators, and the addition of commercially available ChIP-seq kits will streamline the process. The inclusion of ChIP-seq data into foodborne pathogen research will aid in elucidating the mechanisms by which bacteria are capable of persistence, pathogenicity, and virulence. The coupling of RNA-seq and ChIP-seq will provide even more insight into bacterial pathogens.

Proteomics

Cells are constantly integrating external and internal signals to respond appropriately to a changing environment. A large gain of knowledge of the transcriptional regulation dynamics has been achieved through the advent of microarrays and sequencing. Although the mRNA levels in a cell reflect the abundance and activity of their corresponding proteins, proteins are the ones which carry out the cellular functions and in some cases, the transcript abundance is not sufficient to predict protein levels in a steady state or in response to stress. The cellular protein pool is dynamic and complex, including post-transcription processes such as localization, modification and degradation of the proteins themselves. Therefore, proteomics deals with the large-scale study of proteins and their interactions which play an important role in understanding an organism; this area of study is growing rapidly in recent years. Two-dimensional (2D) gel electrophoresis, mass spectrometry (MS) and protein microarrays are major techniques used in proteomics studies.

Two-dimensional sodium-dodecyl-sulfate polyacrylamide gel electrophoresis (2D SDS-PAGE) was introduced to proteomics back in the 1970s [96] and is still in use even though higher sensitivity techniques such as MS are available. Gel-based proteomics can be used for analyzing protein expression, quantitation and post-translational modifications (PTMs). The 2D gel electrophoresis separates proteins based on the amphoteric nature of proteins (isoelectric point) in the first dimension and molecular weight separation in the second dimension. The first dimension separation usually employs isoelectric focusing (IEF, [97]). Basic groups of proteins become positively charged at acidic pH and acidic groups become negatively charged at basic pH, therefore the net charge of proteins will reach zero at their isoelectric point. A gradient of pH is applied to a gel in the first dimension. When an electric potential is also applied across the gel, proteins will migrate along the gel and accumulate at their isoelectric point. For the second dimension separation, SDS electrophoresis is usually used to separate proteins. The pore size of the gel matrix will result in different migration rates of proteins due to their different molecular weights when an electric field is applied. Although a 2D gel can enable visualization and its performance has been improved by a number of technical advancements since its introduction, it has some limitations when used for proteomics studies [97]. The dynamic range is limited to visualize proteins at different concentrations. The workable loading capacity of a gel system affects the sensitivity and dynamic range. In addition, the reproducibility of gels sometimes can be a challenge. In most cases, 2D gel is coupled with MS to obtain more comprehensive proteomics information by a series of processes such as 2D gel separations and spot detection with in-gel digestion and liquid chromatography (LC)/MS processing and analysis.

Mass Spectrometry

Mass spectrometry (MS) has become a key tool in proteomics for the identification and characterization of proteins including primary sequences, PTMs, protein localization, quantification, and protein–protein interactions [98, 99]. MS-based proteomics has achieved tremendous progress during recent decades since the development of protein ionization methods, matrix-assisted laser desorption/ionization (MALDI) and electrospray ionization (ESI), which was recognized by the 2002 Nobel Prize in Chemistry. This section is focused on MS technology relevant to the applied proteomics of foodborne bacterial pathogens

Methodology of Mass Spectrometry Technology

The common procedure for MS-based proteomics includes protein extraction and/or digestion into peptides, protein/peptide separation, protein/peptide ionization, and then mass-to-charge ratio (*m/z*) measurements and recordings. Intact cells can also be directly analyzed using MS [100]. A mass spectrometer consists of three functional units: an ion source that transfers analyte ions into the gas phase, a mass analyzer that separates ions by their *m/z*, and a detector that monitors the number of ions at each *m/z* value. The generic processes of several commonly used ionization methods and mass analyzers are introduced here.

Ionization Methods

MALDI and ESI are the two most commonly used techniques for protein or peptide ionization for mass spectrometric analysis [101–103]. MALDI ionizes samples which are embedded in a saturated solution of a low-mass organic compound (matrix). For bacteria, a microbial colony or protein extracts can be analyzed. The analyte and the matrix co-crystallize and solidify upon drying. A UV laser beam, usually a nitrogen laser beam ($\lambda = 337$ nm), is then focused on a small spot of the matrix-analyte crystalline surface and used to irradiate the sample-matrix crystal. The matrix strongly absorbs around the wavelength of the laser beam. The uptake of energy results in the sublimation of the matrix and sample into the gas phase, and a plume with ions from both the matrix and analyte is formed. ESI directly ionizes liquid samples. Compared to MALDI, ions generated by ESI are stable and not in an excited state, which are less susceptible to rapid decay [104]. A potential is applied to a liquid at low flow rates in a capillary needle and results in the dispersion of the liquid out of the narrow tip of the needle and the formation of charged droplets. The droplets migrate toward a countercharge electrode, driven by the electric field, and skimmed through a heated nozzle or heated curtain gas. Solvent evaporates and solvent molecules leave the droplet as neutral particles, causing the increase of charge density at the surface of the droplets. When the charge density reaches the Rayleigh limit, Coulomb repulsion increases to the same order as the surface tension, and "Coulomb explosion" further tears the large charged droplets into smaller droplets. Repeating this process results in droplets small enough to desorb ions into the ambient gas, and then form "quasi-molecular" ions for mass analysis [97, 101].

Mass Analyzers

A mass analyzer is a chamber with an electrostatic field to separate ions from the source depending on their *m/z* ratio for detection. They are divided into two broad categories [105]: (1) the scanning and ion-beam mass spectrometers, such as time-of-flight (TOF) and quadrupole (Q), and (2) the trapping mass spectrometers, such as ion trap and Fourier transform mass spectrometer (FT-MS) including Orbitrap and Fourier transform ion cyclotron resonance (FTICR). They can be used alone or in tandem. The types commonly used in proteomics research are ion trap, TOF, quadrupole and FTICR [98]. TOF are usually coupled with MALDI for mass measurement of intact peptides, while ESI is mostly coupled with ion-trap (IT) and quadrupole analyzers [98]. See Table 10.3 for a comparison of mass analyzers.

A TOF analyzer is based on accelerating the ions to high kinetic energy by an electrostatic field. Because the electrostatic field is constant, the acceleration results in a specific mass with a specific charge to travel in a different velocity. The ions are then separated along a flight tube by their different velocities. TOF cannot be directly coupled with an ESI source. However, a TOF hybrid with another mass analyzer such as quadrupole or linear ion trap (see in the succeeding paragraphs) is

Table 10.3 Comparison of common mass analyzers used in proteomics

Mass analyzer	Mass resolution[a]	Mass accuracy (ppm)	Sensitivity (mole)	*m/z* range	Tandem MS capability	Ion source
IT	1000–1500	100–1000	Pico	50–2000; 200–4000	MS^n	ESI
LTQ (LIT)	2000	100–500	Femto	50–2000; 200–4000	MS^n	ESI
Q-q-Q	1000	1000–1500	Atto	50–4000	MS/MS	ESI
TOF	10,000–20,000	5–50	Femto	No upper limit	N/A	MALDI
TOF-TOF	10,000–40,000	5–50	Femto	No upper limit	MS/MS	MALDI
Q-q-TOF	10,000–40,000	5–50	Atto	No upper limit	MS/MS	ESI; MALDI
LTQ-FTICR	50,000–800,000	1–2	Femto	50–2000; 200–4000	MS^n	ESI; MALDI
LTQ-Orbitrap	50,000–500,000	<5	Femto	50–2000; 200–4000	MS^n	ESI; MALDI

[a]Mass resolution at full width half maximum
N/A not applicable, *MS^n* multi-stage MS, *MS/MS* two MS in tandem
Reconstructed from Cunsolo et al. [106]

compatible with the EIS source. MALDI coupled with TOF-TOF tandem MS is often used in proteomics. Two TOF sections are connected with a collision cell which undergoes "collision-induced dissociation". This process is used to fragment proteins/peptides based on the collision between a biomolecular ion with a neutral atom or molecule [107]. MALDI TOF-TOF provides a powerful tool for *de novo* sequencing of peptides from in-gel digestion. In addition, TOF is theoretically capable of unlimited mass range analyses, and the collision energy used to produce fragmentation is higher. Thus, TOF-TOF can be used in the analyses of large peptides and intact proteins [106, 107].

A quadrupole mass analyzer uses a strong-focusing alternating gradient field which is applied along the flight-path in the accelerator [108]. In a mass filter which uses two-dimensional (2D) quadrupole field, four circular metal rods paired in opposite directions construct the basic structure of the analyzer. Ions in a specific range of *m/z* are selected by choosing a particular pair of constant voltage (U) and radio-frequency potential (V) applied to the quadrupole structure. The width of the range depends on the ratio of *U/V*, which in principle determines the resolution. Accordingly, ions with different *m/z* can be selected by changing the magnitudes of *U* and *V* [108]. Triple quadrupole mass spectrometer are commonly used in proteome research: ions of a particular *m/z* ratio are selected in the first section which is a 2D quadrupole field (Q), then the ions are fragmented in a collision cell (q), and the fragments are separated in a third quadrupole (Q) section [98].

Quadrupole ion traps, including 3D quadrupole ion trap (IT) and linear quadrupole ion trap (LIT, or linear trap quadrupole, LTQ), share the same principles of field generation as quadrupole mass analyzer. Hybrid quadrupole time-of-fight (QqTOF) mass spectrometer is commonly used in proteomics. Ions from the MALDI ion source can be firstly cooled by a collisional damping interface (q) and then transported through quadrupole (Q) and measured in the TOF section [109], or ions from the ESI ion source can pass through the first and second sections of a triple quadrupole and a reflector TOF (TOF with a reflector to compensate slight differences in kinetic energy of ions) for the measurement [98]. IT is robust, sensitive, and moderately inexpensive, but the resolution and mass accuracy is relatively low [106]. Compared to IT, LTQ has enhanced sensitivity due to the higher injection efficiencies and ion storage capacities [110].

Trapping mass spectrometers include the quadrupole ion traps mentioned above, and also Fourier transform (FT) IT (e.g. Orbitrap and FTICR). Orbitrap and FTICR are based on Fourier transformation of image current of ions to obtain *m/z* spectra [97]. Orbitrap uses electric fields to induce the transient image current, while FTICR employs magnetic fields. Orbitrap is usually connected to a radio frequency-only LTQ which injects pulsed ion beams into the rapidly changing electric field in the Orbitrap [111]. Both FT-MS have the strengths of high mass resolution, accuracy and dynamic range [112]. However, to achieve the high resolution, it usually takes time to pump down the residual gas [107]. The expenses and relatively low peptide-fragmentation efficiency should also be considered for routine use [98].

MS-Based Proteomic Analysis Strategies

MS-based proteomics analysis usually embraces two strategies: (1) bottom-up, which starts with enzymatical or chemical digestion of proteins into peptides which are then analyzed, or (2) top-down, which analyzes intact proteins. The bottom-up strategy is common with high-complexity samples in conjunction with large-scale analyses [105]. Protein samples are extracted from cells and digested to peptides. Then peptides are separated from the sample by chromatography, ionized, and introduced to the mass spectrometer. Usually tandem MS with collision activated dissociation is required. Linear ion trap (LIT), quadrupole-linear ion trap (Q-LIT), triple quadrupole, Q-TOF and IT-TOF are widely used. The MS data identification is based on correlating the experimental data to a database such as Genpept [113]. The drawbacks of using this strategy are that there is only a partial identification of total peptides of a given protein and that there is loss of PTM information [107]. Top-down methods provide advantages such as rapidity, simplified mass spectra, PTM information, protein–protein complex information, protein quantification, and tolerance to contaminants [100, 105]. However, there are limitations such as challenging separation of intact proteins as compared to peptide mixtures. Therefore, LIT-Orbitrap and LIT-ICR are used due to their capability of using larger protein quantities and resulting in higher mass accuracy [105]. For direct microorganism analysis, the top-down strategy can be used to differentiate bacteria at genus, species, and strain levels with the identification of a small portion of the proteome [100]. The expressed sequence tag method and the *de novo* method are used to analyze top-down data [105] The procedure for MS-based food microbial analysis and strategies are shown in Fig. 10.4.

The use of MS-based techniques in food safety need to address the concerns of detection, identification, and quantification of food pathogens and a variety of virulence factors that the pathogens excrete into the food and cell surfaces. Proteomic studies based on MS have been developed for bacterial profiling for distinguishing different species and strains, identifying toxins and pathogenicity determinants such as the synthesis of proteins which correlate with virulence under various conditions, and the interaction between pathogens and their hosts to cause disease [106, 114].

Applications of Mass Spectrometry

Identification of Bacterial Pathogens

MALDI-MS provides a rapid and simplified method to directly analyze, identify and distinguish pathogenic and nonpathogenic bacteria based on the detection of high- and low-molecular-weight proteins, called the "fingerprint" of a specific genus, species or strain [100, 115, 116]. Mass spectra can be obtained from unknown

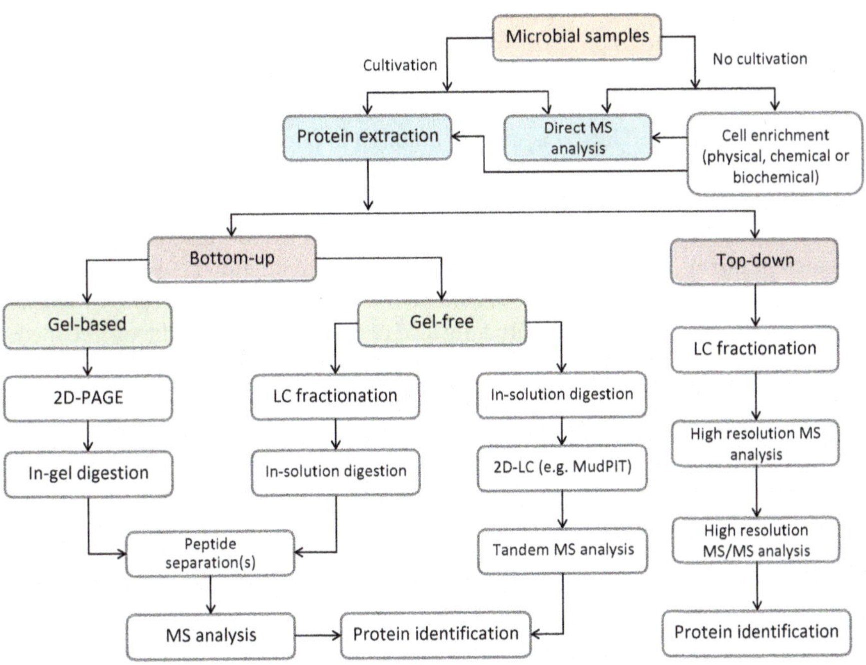

Fig. 10.4 Generic experimental flowchart of MS-based analysis of bacterial samples, including a comparison of bottom-up and top-down strategies [100, 106]. MudPIT: multidimensional protein identification technology

bacterial samples and processed with an inactivation/extraction procedure. The spectra reveal some major peptides and proteins, and have the potential to differentiate pathogens based on the profiling spectra containing a series of peaks [115]. By comparing the mass spectra with reference libraries, the unknown bacteria can be identified as well [117]. For example, Barbuddhe et al. [115] utilized the chemistry of ribosomal proteins to identify *Listeria* species via MALDI-TOF. This procedure was used to differentiate pathogenic and nonpathogenic species. In addition, *L. monocytogenes* isolates were separated up to the level of clonal lineages. The studies also indicate the potential for tracking foodborne pathogen outbreaks and facilitating epidemiological studies.

Quantitation of Proteins of Interest

In addition to the qualitative analysis of the proteome, the abundance, distribution, and stoichiometry are also very important aspects to unveil functional information such as enzymatic reactions and signaling pathways which depend on proteins. The measurement of protein concentrations associated with different states, such as responses to environmental changes, are usually performed in quantitative proteomics. Since MS is highly dependent on the separation of proteins and peptides prior to mass analysis to obtain unambiguous identification, tandem MS coupled with liquid chromatography (LC-MS/MS) has especially become the major technique for proteomics research. However, even the peptides from the same protein may differ in ion intensities due to charge state, peptide length, amino acid composition, or PTM [118]. Thus, it is usually required to compare each peptide between experiments. The relative quantitative methods are used to compare two or more samples by using either stable isotope labeling approaches or label-free approaches. Stable-isotope tags are introduced to proteins by metabolic labelling, enzymatic transfer, or chemical reaction [98]. Based on the differentiation between the heavy/light isotope pairs in the mass spectrometer, the relative-abundance ratio of labeled heavy/light peptide pairs is measured. Instead of using isotope tags, label-free methods use spectrum counting methods or peptide ion intensity methods to determine the abundance of proteins [105, 118]. The spectrum counting methods use the total number of fragmentation spectra that map to peptides of a given protein as a quantitative measure. Using the peptide ion intensity to compare two or more samples, a method called "protein correlation profiling" is used to align the total ion chromatograms of different samples [119]. While the relative quantification methods are applied for a large number of proteins in mixture, absolute quantification is also commonly used, by adding a known quantity of stable isotope-labeled standard peptide to samples, which determines the quantity of one or a few particular proteins [120]. An approach of constructing a synthetic gene encoding a concatenation of tryptic standard peptides, which provides multiple peptides of a target protein or quantification standards for a group of proteins upon digestion, has been developed for absolute quantification [121]. The MS-based quantification has been used to obtain protein profiles of bacteria for studying the response of bacteria to the growth environment, which is important in controlling food processing conditions. The molecular basis of *Cronobacter* sp adaption to heat and cold-stress was investigated with one *C. turicensis* isolate cultured at different temperatures [122]. iTRAQ (isobaric tags for relative and absolute quantification)-labelled whole cells and secreted proteins were identified and quantified by 2-D LC-MALDI-TOF/TOF MS. The study reported the changes in protein expressions, including various potential virulence factors, under different growth temperatures which might explain the high infection potential of *C. turicensis*. Liu et al. [123] studied the proteome of *Campylobacter jejuni* at different times after the infection of cultured mammalian cells using quantitative LC-MS/MS (LTQ) analysis. The analysis indicated a significant metabolic downshift and change in its respiration mode, and provided potential basis for the future development of antimicrobial strategies which target the metabolic pathways of the pathogen [123].

Shotgun Proteomics

Due to the limitations of 2D-polyacrylamide gel electrophoresis (PAGE), the development of alternative approaches for the separation of complex mixtures has attracted interest. This has led to the emergence of a new MS-based technique called shotgun proteomics. Shotgun proteomics techniques are commonly used in quantitative studies. The basis of the procedure is the digestion of a mixture of proteins, 2D chromatography-based separation and tandem MS. The MS data is then analyzed by computational programs. The 2D chromatography for peptide separation uses a combination of chromatographic techniques, such as the most popular multidimensional protein identification technology (MudPIT) which consists of a strong cation exchange (SCX) and a reversed phase (RP) stationary phase packed together. The MudPIT is loaded with a peptide mixture and placed in line between an HPLC and tandem MS. The main advantages of MudPIT over 2D-PAGE lie in its improvements in the detection and identification of membrane proteins and low abundance proteins [124]. Alternative configurations such as affinity chromatography (AC)/RP, isoelectric focusing (IEF)/RP, and capillary electrophoresis (CE)/RP were also investigated [125]. GeLC-MS/MS, where proteins are firstly separated on an SDS-PAGE gel followed by the gel being sliced at specific intervals for analysis, is also available. After the separation by 2D chromatography, peptides are eluted directly into tandem MS. Shotgun proteomics are employed in foodborne bacterial studies. For example, survival mechanisms of *E. coli* O157:H7 were investigated using shotgun proteomics-based analysis using isobaric tags for relative and absolute quantitation [126]. MudPIT and LTQ-Orbitrap tandem MS-based shotgun proteomics were used to reveal the response of *L. monocytogenes* to the extracellular pH environment [127].

Protein–Protein Interactions

Another major application of MS-based proteomics is the analysis of protein interactions. Studying how foodborne pathogens interact with a food matrix or with their host is crucial for developing food processing and antimicrobial strategies. Protein–protein interactions are important in the pathogenic processes of bacteria such as molecular recognition, adherence to host cells, and the regulation of one protein upon another. MS-based proteomics has become a powerful tool for the identification of interacting protein complexes and for interaction profiling. It provides merits such as the capabilities of analyzing interactions taking place in the native environment and in cellular locations, using fully processed and modified proteins as affinity reagents, and isolating and analyzing multicomponent complexes in a single operation [98]. Usually the MS-based protein interaction analysis consists of bait presentation, affinity purification of the complex, and the analysis of bound proteins. The tandem affinity purification (TAP)-MS is one of the most popular methods used for purifying protein complexes from their natural environment for interaction analysis studies. The method uses a fusion protein (bait) including a calmodulin-binding tag in series with an immunoglobulin-binding tag of protein A, where the two tags are connected by a sequence which can be cleaved by tobacco etch virus (TEV) protease [128, 129]. The TAP-fusion is expressed in the host cells, and a protein complex assembles under physiological conditions, with one protein being cloned and tagged. The fusion protein along with associated interacting proteins are retrieved from the cells and purified by reactions between the two tags and immunoglobulin resin and calmodulin resin sequentially. Between the two affinity purifications TEV proteases cleaves the sequence between the two tags. After the two purification steps, the elution including the bait and its interacting proteins can be analyzed by MS. Butland et al. [130] reported the identification of an *E. coli* protein interaction by MS using the TAP procedure. Gel-based peptide mass fingerprinting using MALDI-TOF was employed, and gel-free shotgun sequencing (LC-MS) was used to identify small and lower-abundance proteins. Burnaevskiy et al. [131] studied the *Shigella flexneri* virulence factor, invasion plasmid antigen J (IpaJ), which causes the demyristoylation of human proteins that regulate cargo transport through the Golgi apparatus. The host protein containing a Strep affinity tag and tandem MS were used to observe the cleavage of N-myristoylated proteins which recognized this pathogenic mechanism of *Shigella.*

Protein Modifications

PTMs are important for cellular processes such as the biological functions of proteins, cellular localization of proteins, and protein complex formations [132]. These modifications can change the molecular weight and fundamental physical properties of the proteins. Some examples are phosphorylation (+80 Da), sulfation (+80 Da), nitration (+45 Da), O-glycosylation (>203 Da), and acylation (>200 Da) [133]. MS can be employed to determine the type and site of modifications by comparing the measured mass and fragmentation spectra with a database search [98]. This analysis sometimes can be a challenge to MS due to the mass shift in the peptide molecular weight, the abundance of the modified peptide, the stability of the modification using MS analysis and the sensitivity affected by the modification [132]. Non-gel purification methods are usually preferred to minimize the losses in sequence coverage for modification-site determination, such as ESI-LC-MS/MS. Gel-based separation coupled with LC-MALDI-MS is also used in some studies. Four foodborne bacterial pathogens, *C. jejuni*, *Helicobacter pylori*, *Aeromonas caviae*, and *L. monocytogenes*, were studied by a top-down MS approach for characterizing their protein glycosylation using Q-TOF. The flagella of the four pathogens, which are important for motility, were analyzed and a significant diversity of glycan residues were found on certain flagella proteins [134]. Macek et al. [135] used a gel-free method of HPLC coupled with LTQ-FT or LTQ-Orbitrap to identify 103 phosphopeptides from 78 *Bacillus subtilis* proteins and 78 phosphorylation sites. The phosphorylation on serine, threonine, and tyrosine as a key regulatory PTM in bacteria was investigated. Myriad databases, software and tools used to interpret the PTMs are currently available, which provide valuable resources for PTM research [136].

Limitations and Challenges

The major challenges of MS-based proteomics are to achieve comprehensive, reproducible and quantitative description of proteomes with reasonable throughput [137]. MS-based proteomics datasets tend to be biased against lower abundance proteins which results in incomplete proteome coverage [118]. In addition, although studies on protein interactions have matured, coverage is still quite low especially with interactions between proteins and other molecules and cannot always fully describe dynamic protein networks [128, 137]. This is greatly attributed to the challenge of obtaining and isolating affinity molecules for large scale analysis. Also, studies of some functional proteins such as membrane proteins requires an artificial environment that mimics the native membrane, therefore some techniques used in MS can be detrimental to protein activity and stability [138]. Although MS data analysis is not discussed in this chapter, it should be noted here that enormous amounts of data are generated by various techniques and experimental designs bring additional complexity; this requires further development of appropriate statistical approaches, databases and tools to reduce the time and effort required to provide meaningful interpretations of the results.

Protein Microarray

Protein microarray provides a platform for analyzing thousands of proteins simultaneously, which allows for its robust application in proteomics. There are two main categories of protein microarrays [139]: (1) protein detection, where specific protein-capture reagents such as antibodies and aptamers are immobilized on the arrays, which can specifically recognize particular proteins in complex mixtures, and detect the abundance and PTMs of the proteins; and (2) protein function determination, where purified proteins, protein domains or functional peptides are immobilized on the arrays, and their interactions with other proteins, small molecules and substrates for enzymes are studied.

Methodology of Protein Microarray Technology

For protein detection arrays, there are several common configurations (Fig. 10.5): (1) target direct labeling, where the target proteins are labeled with tags such as fluorescent dyes, radioisotopes, and even nanoparticles which have emerged in recent years. The target proteins can be captured by the immobilized binding molecules, and detected through the signals from the tags. The readout is usually based on fluorescence, chemiluminescence, mass spectrometry, radioactivity or electrochemistry; (2) label-free, where the detection is based on inherent properties of the target proteins; (3) sandwich assay, where the target proteins are captured by their specific binding molecules, followed by the attachment of a secondary labeled molecule (e.g. antibody) which binds the target; and (4) reverse phase protein blot, where cell lysates or blood fluid are immobilized on microarray spots and then a labeled detection molecule (e.g. antibody) is incubated with the complex mixtures to detect the target protein [140].

For functional microarrays, target-purified proteins, protein domains or functional peptides can be spotted on arrays through various methods (Fig. 10.6): (1) purified proteins immobilized on chemically functionalized glass slides; (2) glutathione-S-transferase (GST)-$(His)_6$ tagged fusion proteins which can be immobilized on Ni-coated slides; (3) Nucleic acid-programmable protein array (NAPPA), where plasmid DNAs are spotted on the array instead of purified proteins. The proteins corresponding to the plasmid DNAs are expressed in situ just prior to experimentation with an epitope tag which can be captured by a capture reagent immobilized on the array with the DNAs; and (4) multiple spotting technique (MIST), where PCR products are immobilized, and proteins are expressed *in situ* [140, 141].

Applications of Protein Microarray

Quantitative Proteomics

Since protein microarrays have many affinity binding reagents immobilized at high spatial density and each reagent captures its target protein, the captured target protein is concentrated only in a small area on a microspot. The specific capture of the target protein also ensures dose-dependent signals. Therefore, the captured proteins

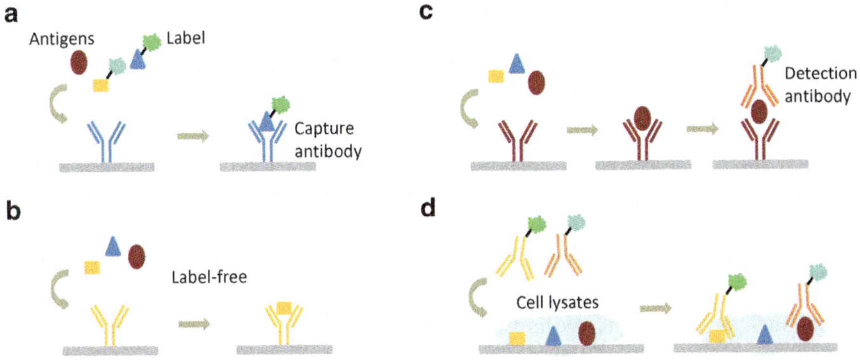

Fig. 10.5 Common formats of protein detection microarrays. (**a**) target direct labeling method; (**b**) label-free detection method; (**c**) sandwich immunoassay; and (**d**) reverse phase protein blot

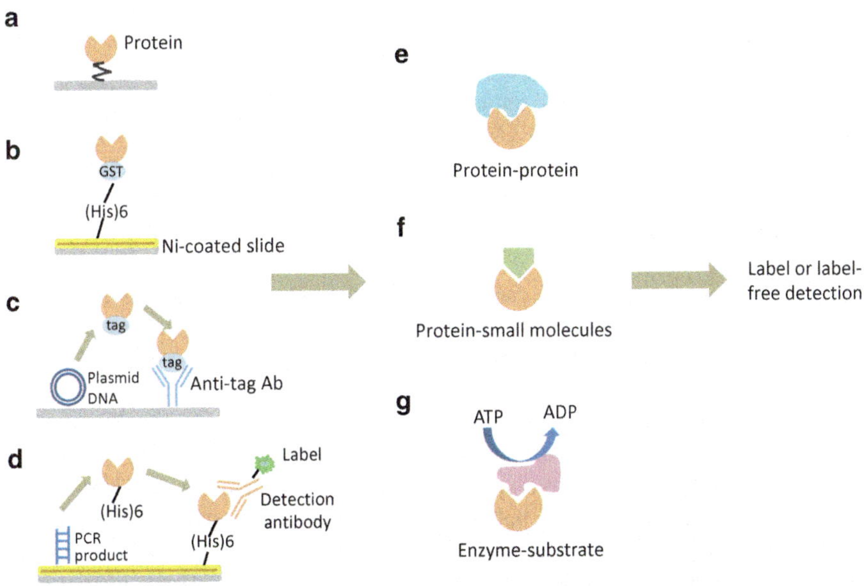

Fig. 10.6 Common formats of protein function microarrays. Proteins are spotted by (**a**) chemical immobilization method; (**b**) fusion protein method; (**c**) nucleic acid-programmable protein array (NAPPA) and (**d**) multiple spotting technique (MIST)-based protein expression from PCR products. Functional microarrays can be used to study (**e**) protein–protein interactions; (**f**) protein-small molecule interactions; and (**g**) interactions between enzymes and substrates

can be detected and quantified with high signal intensity and low signal-to-noise ratios [142, 143]. Moreover, the capability of analyzing thousands of different parameters with robustness allows the application of protein microarrays in the proteomics studies of foodborne pathogens. Depending on the label used, fluorescence, chemiluminescence, mass spectrometry, radioactivity or electrochemistry signals can be associated with the quantity of target proteins. Danckert et al. [144] used protein microarrays to identify and study novel immunodominant proteins of *S.* Enteritidis. The authors identified proteins such as SEN2278 and SEN4030 to be involved in immunogenicity and may be potential candidates for *S. enterica*-specific diagnosis. Gehring et al. [145] used a sandwich protein microarray for different Shiga toxins for detection and quantitation of *E. coli* strains. The assay can be used for detection of the pathogen in clinical and food samples.

For label-free detection, where the measurement is based on inherent properties of the target proteins such as mass and dielectric properties [140], optical techniques such as surface plasmon resonance (SPR) and ellipsometry and microscopic techniques such as atomic force microscopy have been integrated with protein microarrays. In recent years, nanotechnology also finds its way for integration into protein microarrays. Quantum dots and gold nanoparticles are used as labels, and electrical property changes in functionalized nanowires due to the binding of a target protein have been measured for label free detection [140, 146].

PTMs can also be identified by protein microarrays. Antibody-based western blots have been used to identify PTMs such as tyrosine phosphorylation. However, some PTMs with small-sized structural motifs may not be recognized due to the difficulty in generating specific antibodies [147]. Enzymes specific to PTMs have also been employed to identify PTM sequence motifs [148].

Functional Proteomics

Functional protein microarrays can have various configurations depending on the applications. The advantages of low reagent consumption, rapid result readouts and easily controlled experimental conditions allow microarrays to be useful in rapid screening of large numbers of proteins for their biochemical activities and interactions between proteins and between proteins and other molecules such as nucleic acids, lipids and small organic compounds [142, 149]. Using functional protein microarrays to study protein interactions, such as the mechanisms of bacterial drug resistance and the host immune response towards bacterial pathogens, has contributed to the design of therapeutic strategies, drug development, and disease diagnosis. For example, the interactions between *E. coli* proteins which mediate the adhesion of the bacteria to epithelial cells and their ligand were studied by using protein arrays immobilized with the ligand. Inhibitors of *E. coli* adhesion were also screened and quantified by immobilizing the inhibitors on the arrays [150]. The entire proteome of *E. coli* K12 was printed on a protein array to screen and identify biomarkers for inflammatory bowel disease. Interactions between *E. coli* proteins and human serum antibodies were analyzed to differentiate responses between healthy controls, patients with Crohn disease and ulcerative colitis, and new sets of biomarkers to diagnose these ailments were identified [151]. Liu et al. [152] also studied *E. coli* using a protein functional microarray. The authors created an entire proteome microarray to determine the functional interactions of CobB, a deacetylase. Acetylation and deacetylation plays an important role in many biological processes. CobB was determined to interact strongly with proteins such as AccC, acetyl-CoA carboxyltransferase.

Limitations and Challenges

Compared to 2D gel electrophoresis and mass spectrometry, which can be used to analyze both known and unknown proteins, protein microarray is better suited for analyzing a particular target set of known proteins [142]. Accordingly, protein microarrays are largely dependent on binding molecules such as antibodies and the affinity of proteins. For some protein targets, high-affinity and high-specificity antibodies are still not available. There are a limited number of studies on the cross-reactivity of antibodies with other cellular proteins, which hinders the quantitative analysis of cellular lysates [142]. Sample preparation and spotting also affects the accuracy and reproducibility of the quantification. For whole-proteome microarrays, not only is there the challenge of the isolation of a large number of functional proteins, but also the challenge of changing structures, functions and abundance of target proteins which makes downstream analysis extremely complicated. Due to the complexity of the cellular lysates, more research has been done with cell secretions rather than cellular lysates.

Conclusions and Future Perspectives

Proteomics technology, including MS and protein microarray, is important in the study of foodborne bacterial pathogens. This technology has changed the way researchers identify and study proteins. Using MS, studies have determined sequences of important proteins, PTMs and various other modifications, localization of proteins within complex regulatory networks, and protein–protein interactions. MS has also been an important tool for the evaluation and validation of existing genomic annotations. In the future, strain-typing with MS technology should provide a more rapid and robust method for clinical microbiology. Strain identification and subtyping of bacterial pathogens, including antibiotic resistance profiles, will be more efficient. New techniques of ionization and fragmentation might enhance the availability of MS analysis for previous undetected peptides. The development of other non-MS proteomic approaches will provide a complementary view of whole proteomics technologies. It is expected that advances in data mining methods, bioinformatics tools, and protein tagging and affinity purification techniques will emerge for future proteomics research. The study of cellular dynamics at the protein level still has much to achieve in the coming years. Combining genomic, transcriptomic and proteomic data from the same biological system will significantly increase our understanding of complex biological process.

References

1. Nagalakshmi U, Wang Z, Waern K, Shou C, Raha D, Gerstein M, Snyder M. The transcriptional landscape of the yeast genome defined by RNA sequencing. Science. 2008;320(5881):1344–9. doi:10.1126/science.1158441.
2. Wilhelm BT, Marguerat S, Watt S, Schubert F, Wood V, Goodhead I, et al. Dynamic repertoire of a eukaryotic transcriptome surveyed at single-nucleotide resolution. Nature. 2008;453(7199):1239–43. doi:10.1038/nature07002.
3. Mao C, Evans C, Jensen RV, Sobral BW. Identification of new genes in *Sinorhizobium meliloti* using the Genome Sequencer FLX system. BMC Microbiol. 2008;8:72. doi:10.1186/1471-2180-8-72.
4. Giannoukos G, Ciulla DM, Huang K, Haas BJ, Izard J, Levin JZ, et al. Efficient and robust RNA-seq process for cultured bacteria and complex community transcriptomes. Genome Biol. 2012;13(3):R23. doi:10.1186/gb-2012-13-3-r23.
5. Liu JM, Livny J, Lawrence MS, Kimball MD, Waldor MK, Camilli A. Experimental discovery of sRNAs in *Vibrio cholerae* by direct cloning, 5S/tRNA depletion and parallel sequencing. Nucleic Acids Res. 2009;37(6), e46. doi:10.1093/nar/gkp080.
6. Yi H, Cho YJ, Won S, Lee JE, Jin Yu H, Kim S, et al. Duplex-specific nuclease efficiently removes rRNA for prokaryotic RNA-seq. Nucleic Acids Res. 2011;39(20), e140. doi:10.1093/nar/gkr617.
7. Croucher NJ, Fookes MC, Perkins TT, Turner DJ, Marguerat SB, Keane T, et al. A simple method for directional transcriptome sequencing using Illumina technology. Nucleic Acids Res. 2009;37(22), e148. doi:10.1093/nar/gkp811.
8. Perkins TT, Kingsley RA, Fookes MC, Gardner PP, James KD, Yu L, et al. A strand-specific RNA-Seq analysis of the transcriptome of the typhoid bacillus *Salmonella* typhi. PLoS Genet. 2009;5(7), e1000569. doi:10.1371/journal.pgen.1000569.
9. Metzler ML. Sequencing technologies- the next generation. Nat Rev Genet. 2010;11:31–46.
10. Bentley DR, Balasubramanian S, Swerdlow HP, Smith GP, Milton J, Brown CG, et al. Accurate whole human genome sequencing using reversible terminator chemistry. Nature. 2008;456(7218):53–9. doi:10.1038/nature07517.
11. Mortazavi A, Williams BA, McCue K, Schaeffer L, Wold B. Mapping and quantifying mammalian transcriptomes by RNA-Seq. Nat Methods. 2008;5(7):621–8. doi:10.1038/nmeth.1226.
12. Pepke S, Wold B, Mortazavi A. Computation for ChIP-seq and RNA-seq studies. Nat Methods. 2009;6(11 Suppl):S22–32. doi:10.1038/nmeth.1371.
13. Yoder-Himes DR, Chain PS, Zhu Y, Wurtzel O, Rubin EM, Tiedje JM, Sorek R. Mapping the *Burkholderia cenocepacia* niche response via high-throughput sequencing. Proc Natl Acad Sci U S A. 2009;106(10):3976–81. doi:10.1073/pnas.0813403106.
14. Babraham Institute. FastQC. 2010. www.bioinformatics.babraham.ac.uk/projects/fastqc.
15. Qu W, Hashimoto S, Morishita S. Efficient frequency-based de novo short-read clustering for error trimming in next-generation sequencing. Genome Res. 2009;19(7):1309–15. doi:10.1101/gr.089151.108.
16. DeLuca DS, Levin JZ, Sivachenko A, Fennell T, Nazaire MD, Williams C, et al. RNA-SeQC: RNA-seq metrics for quality control and process optimization. Bioinformatics. 2012;28(11):1530–2. doi:10.1093/bioinformatics/bts196.
17. Wang L, Wang S, Li W. RSeQC: quality control of RNA-seq experiments. Bioinformatics. 2012;28(16):2184–5. doi:10.1093/bioinformatics/bts356.
18. Morgan M, Anders S, Lawrence M, Aboyoun P, Pagès H, Gentleman R. ShortRead: a bioconductor package for input, quality assessment and exploration of high-throughput sequence data. Bioinformatics. 2009;25(19):2607–8. doi:10.1093/bioinformatics/btp450.
19. Bolger AM, Lohse M, Usadel B. Trimmomatic: a flexible trimmer for Illumina sequence data. Bioinformatics. 2014;30(15):2114–20. doi:10.1093/bioinformatics/btu170.

20. Magoč T, Salzberg SL. FLASH: fast length adjustment of short reads to improve genome assemblies. Bioinformatics. 2011;27(21):2957–63. doi:10.1093/bioinformatics/btr507.
21. Langmead B, Trapnell C, Pop M, Salzberg SL. Ultrafast and memory-efficient alignment of short DNA sequences to the human genome. Genome Biol. 2009;10(3):R25. doi:10.1186/gb-2009-10-3-r25.
22. Li H, Durbin R. Fast and accurate long-read alignment with Burrows-Wheeler transform. Bioinformatics. 2010;26(5):589–95. doi:10.1093/bioinformatics/btp698.
23. Clement NL, Snell Q, Clement MJ, Hollenhorst PC, Purwar J, Graves BJ, et al. The GNUMAP algorithm: unbiased probabilistic mapping of oligonucleotides from next-generation sequencing. Bioinformatics. 2010;26(1):38–45. doi:10.1093/bioinformatics/btp614.
24. Weese D, Emde AK, Rausch T, Döring A, Reinert K. RazerS: fast read mapping with sensitivity control. Genome Res. 2009;19(9):1646–54. doi:10.1101/gr.088823.108.
25. Trapnell C, Williams BA, Pertea G, Mortazavi A, Kwan G, van Baren MJ, et al. Transcript assembly and quantification by RNA-Seq reveals unannotated transcripts and isoform switching during cell differentiation. Nat Biotechnol. 2010;28(5):511–5. doi:10.1038/nbt.1621.
26. Anders S, Huber W. Differential expression analysis for sequence count data. Genome Biol. 2010;11(10):R106. doi:10.1186/gb-2010-11-10-r106.
27. Magoc T, Wood D, Salzberg SL. EDGE-pro: estimated degree of gene expression in prokaryotic genomes. Evol Bioinform Online. 2013;9:127–36. doi:10.4137/EBO.S11250.
28. Wang Z, Gerstein M, Snyder M. RNA-Seq: a revolutionary tool for transcriptomics. Nat Rev Genet. 2009;10(1):57–63. doi:10.1038/nrg2484.
29. Rutherford K, Parkhill J, Crook J, Horsnell T, Rice P, Rajandream MA, Barrell B. Artemis: sequence visualization and annotation. Bioinformatics. 2000;16(10):944–5.
30. Robinson JT, Thorvaldsdóttir H, Winckler W, Guttman M, Lander ES, Getz G, Mesirov JP. Integrative genomics viewer. Nat Biotechnol. 2011;29(1):24–6. doi:10.1038/nbt.1754.
31. Abeel T, Van Parys T, Saeys Y, Galagan J, Van de Peer Y. GenomeView: a next-generation genome browser. Nucleic Acids Res. 2012;40(2), e12. doi:10.1093/nar/gkr995.
32. Bae D, Crowley MR, Wang C. Transcriptome analysis of *Listeria* monocytogenes grown on a ready-to-eat meat matrix. J Food Prot. 2011;74(7):1104–11. doi:10.4315/0362-028X.JFP-10-508.
33. Bergholz TM, Vanaja SK, Whittam TS. Gene expression induced in *Escherichia coli* O157:H7 upon exposure to model apple juice. Appl Environ Microbiol. 2009;75(11):3542–53. doi:10.1128/AEM.02841-08.
34. Cretenet M, Laroute V, Ulvé V, Jeanson S, Nouaille S, Even S, et al. Dynamic analysis of the *Lactococcus lactis* transcriptome in cheeses made from milk concentrated by ultrafiltration reveals multiple strategies of adaptation to stresses. Appl Environ Microbiol. 2011;77(1):247–57. doi:10.1128/AEM.01174-10.
35. Fratamico PM, Wang S, Yan X, Zhang W, Li Y. Differential gene expression of *E. coli* O157:H7 in ground beef extract compared to tryptic soy broth. J Food Sci. 2011;76(1):M79–87. doi:10.1111/j.1750-3841.2010.01952.x.
36. Liu Y, Ream A. Gene expression profiling of *Listeria* monocytogenes strain F2365 during growth in ultrahigh-temperature-processed skim milk. Appl Environ Microbiol. 2008;74(22):6859–66. doi:10.1128/AEM.00356-08.
37. Makhzami S, Quénée P, Akary E, Bach C, Aigle M, Delacroix-Buchet A, et al. In situ gene expression in cheese matrices: application to a set of enterococcal genes. J Microbiol Methods. 2008;75(3):485–90. doi:10.1016/j.mimet.2008.07.025.
38. Rantsiou K, Greppi A, Garosi M, Acquadro A, Mataragas M, Cocolin L. Strain dependent expression of stress response and virulence genes of *Listeria* monocytogenes in meat juices as determined by microarray. Int J Food Microbiol. 2012;152(3):116–22. doi:10.1016/j.ijfoodmicro.2011.08.009.
39. Sirsat SA, Muthaiyan A, Ricke SC. Optimization of the RNA extraction method for transcriptome studies of *Salmonella* inoculated on commercial raw chicken breast samples. BMC Res Notes. 2011;4:60. doi:10.1186/1756-0500-4-60.

40. Oliver HF, Orsi RH, Ponnala L, Keich U, Wang W, Sun Q, et al. Deep RNA sequencing of *L. monocytogenes* reveals overlapping and extensive stationary phase and sigma B-dependent transcriptomes, including multiple highly transcribed noncoding RNAs. BMC Genomics. 2009;10:641. doi:10.1186/1471-2164-10-641.
41. Deng X, Li Z, Zhang W. Transcriptome sequencing of *Salmonella enterica* serovar Enteritidis under desiccation and starvation stress in peanut oil. Food Microbiol. 2012;30(1):311–5. doi:10.1016/j.fm.2011.11.001.
42. Landick R, Vaughn V, Lau ET, VanBogelen RA, Erickson JW, Neidhardt FC. Nucleotide sequence of the heat shock regulatory gene of *E. coli* suggests its protein product may be a transcription factor. Cell. 1984;38(1):175–82.
43. Hiratsu K, Amemura M, Nashimoto H, Shinagawa H, Makino K. The rpoE gene of *Escherichia coli*, which encodes sigma E, is essential for bacterial growth at high temperature. J Bacteriol. 1995;177(10):2918–22.
44. Brankatschk K, Kamber T, Pothier JF, Duffy B, Smits TH. Transcriptional profile of *Salmonella enterica* subsp. *enterica* serovar Weltevreden during alfalfa sprout colonization. Microb Biotechnol. 2014;7(6):528–44. doi:10.1111/1751-7915.12104.
45. Feng S, Eucker TP, Holly MK, Konkel ME, Lu X, Wang S. Investigating the responses of *Cronobacter sakazakii* to garlic-derived organosulfur compounds: a systematic study of pathogenic-bacterium injury by use of high-throughput whole-transcriptome sequencing and confocal micro-raman spectroscopy. Appl Environ Microbiol. 2014;80(3):959–71. doi:10.1128/AEM.03460-13.
46. Friedemann M. Epidemiology of invasive neonatal *Cronobacter* (*Enterobacter sakazakii*) infections. Eur J Clin Microbiol Infect Dis. 2009;28(11):1297–304. doi:10.1007/s10096-009-0779-4.
47. Fox EM, Leonard N, Jordan K. Physiological and transcriptional characterization of persistent and nonpersistent *Listeria* monocytogenes isolates. Appl Environ Microbiol. 2011;77(18):6559–69. doi:10.1128/AEM.05529-11.
48. Casey A, Fox EM, Schmitz-Esser S, Coffey A, McAuliffe O, Jordan K. Transcriptome analysis of *Listeria* monocytogenes exposed to biocide stress reveals a multi-system response involving cell wall synthesis, sugar uptake, and motility. Front Microbiol. 2014;5:68. doi:10.3389/fmicb.2014.00068.
49. Taveirne ME, Theriot CM, Livny J, DiRita VJ. The complete *Campylobacter jejuni* transcriptome during colonization of a natural host determined by RNAseq. PLoS One. 2013;8(8), e73586. doi:10.1371/journal.pone.0073586.
50. Jin W, Ibeagha-Awemu EM, Liang G, Beaudoin F, Zhao X, Guan IL. Transcriptome microRNA profiling of bovine mammary epithelial cells challenged with *Escherichia coli* or *Staphylococcus aureus* bacteria reveals pathogen directed microRNA expression profiles. BMC Genomics. 2014;15:181. doi:10.1186/1471-2164-15-181.
51. Kröger C, Colgan A, Srikumar S, Händler K, Sivasankaran SK, Hammarlöf DL, et al. An infection-relevant transcriptomic compendium for *Salmonella enterica* Serovar Typhimurium. Cell Host Microbe. 2013;14(6):683–95. doi:10.1016/j.chom.2013.11.010.
52. Mraheil MA, Billion A, Mohamed W, Mukherjee K, Kuenne C, Pischimarov J, et al. The intracellular sRNA transcriptome of *Listeria* monocytogenes during growth in macrophages. Nucleic Acids Res. 2011;39(10):4235–48. doi:10.1093/nar/gkr033.
53. Sorek R, Cossart P. Prokaryotic transcriptomics: a new view on regulation, physiology and pathogenicity. Nat Rev Genet. 2010;11(1):9–16. doi:10.1038/nrg2695.
54. Cho BK, Zengler K, Qiu Y, Park YS, Knight EM, Barrett CL, et al. The transcription unit architecture of the *Escherichia coli* genome. Nat Biotechnol. 2009;27(11):1043–9. doi:10.1038/nbt.1582.
55. Sharma CM, Hoffmann S, Darfeuille F, Reignier J, Findeiss S, Sittka A, et al. The primary transcriptome of the major human pathogen *Helicobacter pylori*. Nature. 2010;464(7286):250–5. doi:10.1038/nature08756.

56. Levin JZ, Yassour M, Adiconis X, Nusbaum C, Thompson DA, Friedman N, et al. Comprehensive comparative analysis of strand-specific RNA sequencing methods. Nat Methods. 2010;7(9):709–15. doi:10.1038/nmeth.1491.
57. Parkhomchuk D, Borodina T, Amstislavskiy V, Banaru M, Hallen L, Krobitsch S, et al. Transcriptome analysis by strand-specific sequencing of complementary DNA. Nucleic Acids Res. 2009;37(18), e123. doi:10.1093/nar/gkp596.
58. Williams AG, Thomas S, Wyman SK, Holloway AK. RNA-seq data: challenges in and recommendations for experimental design and analysis. Curr Protoc Hum Genet. 2014;83:11.13.11–20. doi:10.1002/0471142905.hg1113s83.
59. Ben-Yehuda S, Fujita M, Liu XS, Gorbatyuk B, Skoko D, Yan J, et al. Defining a centromere-like element in *Bacillus subtilis* by identifying the binding sites for the chromosome-anchoring protein RacA. Mol Cell. 2005;17(6):773–82. doi:10.1016/j.molcel.2005.02.023.
60. Ben-Yehuda S, Rudner DZ, Losick R. RacA, a bacterial protein that anchors chromosomes to the cell poles. Science. 2003;299(5606):532–6. doi:10.1126/science.1079914.
61. Lucchini S, McDermott P, Thompson A, Hinton JC. The H-NS-like protein StpA represses the RpoS (sigma 38) regulon during exponential growth of *Salmonella* Typhimurium. Mol Microbiol. 2009;74(5):1169–86. doi:10.1111/j.1365-2958.2009.06929.x.
62. Navarre WW, Porwollik S, Wang Y, McClelland M, Rosen H, Libby SJ, Fang FC. Selective silencing of foreign DNA with low GC content by the H-NS protein in *Salmonella*. Science. 2006;313(5784):236–8. doi:10.1126/science.1128794.
63. Grainger DC, Hurd D, Goldberg MD, Busby SJ. Association of nucleoid proteins with coding and non-coding segments of the *Escherichia coli* genome. Nucleic Acids Res. 2006;34(16):4642–52. doi:10.1093/nar/gkl542.
64. Grainger DC, Hurd D, Harrison M, Holdstock J, Busby SJ. Studies of the distribution of *Escherichia coli* cAMP-receptor protein and RNA polymerase along the *E. coli* chromosome. Proc Natl Acad Sci U S A. 2005;102(49):17693–8. doi:10.1073/pnas.0506687102.
65. Herring CD, Raffaelle M, Allen TE, Kanin EI, Landick R, Ansari AZ, Palsson B. Immobilization of *Escherichia coli* RNA polymerase and location of binding sites by use of chromatin immunoprecipitation and microarrays. J Bacteriol. 2005;187(17):6166–74. doi:10.1128/JB.187.17.6166-6174.2005.
66. Oshima T, Ishikawa S, Kurokawa K, Aiba H, Ogasawara N. *Escherichia coli* histone-like protein H-NS preferentially binds to horizontally acquired DNA in association with RNA polymerase. DNA Res. 2006;13(4):141–53. doi:10.1093/dnares/dsl009.
67. Reppas NB, Wade JT, Church GM, Struhl K. The transition between transcriptional initiation and elongation in *E. coli* is highly variable and often rate limiting. Mol Cell. 2006;24(5):747–57. doi:10.1016/j.molcel.2006.10.030.
68. Wade JT, Castro Roa D, Grainger DC, Hurd D, Busby SJ, Struhl K, Nudler E. Extensive functional overlap between sigma factors in *Escherichia coli*. Nat Struct Mol Biol. 2006;13(9):806–14. doi:10.1038/nsmb1130.
69. Danielli A, Roncarati D, Delany I, Chiarini V, Rappuoli R, Scarlato V. In vivo dissection of the *Helicobacter pylori* Fur regulatory circuit by genome-wide location analysis. J Bacteriol. 2006;188(13):4654–62. doi:10.1128/JB.00120-06.
70. Molle V, Nakaura Y, Shivers RP, Yamaguchi H, Losick R, Fujita Y, Sonenshein AL. Additional targets of the *Bacillus subtilis* global regulator CodY identified by chromatin immunoprecipitation and genome-wide transcript analysis. J Bacteriol. 2003;185(6):1911–22.
71. Wade JT, Reppas NB, Church GM, Struhl K. Genomic analysis of LexA binding reveals the permissive nature of the *Escherichia coli* genome and identifies unconventional target sites. Genes Dev. 2005;19(21):2619–30. doi:10.1101/gad.1355605.
72. Thijs IM, De Keersmaecker SC, Fadda A, Engelen K, Zhao H, McClelland M, et al. Delineation of the *Salmonella enterica* serovar Typhimurium HilA regulon through genome-wide location and transcript analysis. J Bacteriol. 2007;189(13):4587–96. doi:10.1128/JB.00178-07.

73. Tomljenovic-Berube AM, Mulder DT, Whiteside MD, Brinkman FS, Coombes BK. Identification of the regulatory logic controlling *Salmonella* pathoadaptation by the SsrA-SsrB two-component system. PLoS Genet. 2010;6(3), e1000875. doi:10.1371/journal.pgen.1000875.
74. Hopp TP, Prickett KS, Price VL, Libby RT, March CJ, Cerretti DP, et al. A short polypeptide marker sequence useful for recombinant protein identification and purification. Nat Biotechnol. 1988;6:1204–10.
75. Hochuli E, Bannwarth W, Döbeli H, Gentz R, Stüber D. Genetic approach to facilitate purification of recombinant proteins with a novel metal chelate adsorbent. Nat Biotechnol. 1988;6:1321–5.
76. Zhang Y, Liu T, Meyer CA, Eeckhoute J, Johnson DS, Bernstein BE, et al. Model-based analysis of ChIP-Seq (MACS). Genome Biol. 2008;9(9):R137. doi:10.1186/gb-2008-9-9-r137.
77. Rozowsky J, Euskirchen G, Auerbach RK, Zhang ZD, Gibson T, Bjornson R, et al. PeakSeq enables systematic scoring of ChIP-seq experiments relative to controls. Nat Biotechnol. 2009;27(1):66–75. doi:10.1038/nbt.1518.
78. Fejes AP, Robertson G, Bilenky M, Varhol R, Bainbridge M, Jones SJ. FindPeaks 3.1: a tool for identifying areas of enrichment from massively parallel short-read sequencing technology. Bioinformatics. 2008;24(15):1729–30. doi:10.1093/bioinformatics/btn305.
79. Ji H, Jiang H, Ma W, Johnson DS, Myers RM, Wong WH. An integrated software system for analyzing ChIP-chip and ChIP-seq data. Nat Biotechnol. 2008;26(11):1293–300. doi:10.1038/nbt.1505.
80. Bonocora RP, Fitzgerald DM, Stringer AM, Wade JT. Non-canonical protein-DNA interactions identified by ChIP are not artifacts. BMC Genomics. 2013;14:254. doi:10.1186/1471-2164-14-254.
81. Shimada T, Ishihama A, Busby SJ, Grainger DC. The *Escherichia coli* RutR transcription factor binds at targets within genes as well as intergenic regions. Nucleic Acids Res. 2008;36(12):3950–5. doi:10.1093/nar/gkn339.
82. Garner MM, Revzin A. A gel electrophoresis method for quantifying the binding of proteins to specific DNA regions: application to components of the *Escherichia coli* lactose operon regulatory system. Nucleic Acids Res. 1981;9(13):3047–60.
83. Kadonaga JT, Tjian R. Affinity purification of sequence-specific DNA binding proteins. Proc Natl Acad Sci U S A. 1986;83(16):5889–93.
84. Stringer AM, Currenti S, Bonocora RP, Baranowski C, Petrone BL, Palumbo MJ, Reilly AA, Zhang Z, Erill I, Wade JT. Genome-scale analyses of *Escherichia coli and Salmonella enterica* AraC reveal noncanonical targets and an expanded core regulon. J Bacteriol. 2014;196(3):660–71. doi: 10.1128/JB.01007-13.
85. Davies BW, Bogard RW, Mekalanos JJ. Mapping the regulon of *Vibrio cholerae* ferric uptake regulator expands its known network of gene regulation. Proc Natl Acad Sci U S A. 2011;108(30):12467–72. doi:10.1073/pnas.1107894108.
86. Fitzgerald DM, Bonocora RP, Wade JT. Comprehensive mapping of the *Escherichia coli* flagellar regulatory network. PLoS Genet. 2014;10(10), e1004649. doi:10.1371/journal.pgen.1004649.
87. Wang H, Liu B, Wang Q, Wang L. Genome-wide analysis of the salmonella Fis regulon and its regulatory mechanism on pathogenicity islands. PLoS One. 2013;8(5), e64688. doi:10.1371/journal.pone.0064688.
88. Petrone BL, Stringer AM, Wade JT. Identification of HilD-regulated genes in *Salmonella enterica* serovar Typhimurium. J Bacteriol. 2014;196(5):1094–101. doi:10.1128/JB.01449-13.
89. Reynolds CM, Ribeiro AA, McGrath SC, Cotter RJ, Raetz CR, Trent MS. An outer membrane enzyme encoded by *Salmonella typhimurium* lpxR that removes the 3′-acyloxyacyl moiety of lipid A. J Biol Chem. 2006;281(31):21974–87. doi:10.1074/jbc.M603527200.

90. Kawano M, Manabe T, Kawasaki K. *Salmonella enterica* serovar Typhimurium lipopolysaccharide deacylation enhances its intracellular growth within macrophages. FEBS Lett. 2010;584(1):207–12. doi:10.1016/j.febslet.2009.11.062.
91. Kawasaki K, Teramoto M, Tatsui R, Amamoto S. Lipid A 3′-O-deacylation by *Salmonella* outer membrane enzyme LpxR modulates the ability of lipid A to stimulate Toll-like receptor 4. Biochem Biophys Res Commun. 2012;428(3):343–7. doi:10.1016/j.bbrc.2012.10.054.
92. Perkins TT, Davies MR, Klemm EJ, Rowley G, Wileman T, James K, et al. ChIP-seq and transcriptome analysis of the OmpR regulon of *Salmonella enterica* serovars Typhi and Typhimurium reveals accessory genes implicated in host colonization. Mol Microbiol. 2013;87(3):526–38. doi:10.1111/mmi.12111.
93. Carver T, Harris SR, Berriman M, Parkhill J, McQuillan JA. Artemis: an integrated platform for visualization and analysis of high-throughput sequence-based experimental data. Bioinformatics. 2012;28(4):464–9. doi:10.1093/bioinformatics/btr703.
94. Taylor J, Schenck I, Blankenberg D, Nekrutenko A. Using galaxy to perform large-scale interactive data analyses. Curr Protoc Bioinformatics. 2007; Chapter 10, Unit 10.15. doi:10.1002/0471250953.bi1005s19.
95. Westermann AJ, Gorski SA, Vogel J. Dual RNA-seq of pathogen and host. Nat Rev Microbiol. 2012;10(9):618–30. doi:10.1038/nrmicro2852.
96. O'Farrell PH. High resolution two-dimensional electrophoresis of proteins. J Biol Chem. 1975;250:4007–21.
97. Wright PC, Noirel J, Ow SY, Fazeli A. A review of current proteomics technologies with a survey on their widespread use in reproductive biology investigations. Theriogenology. 2012;77(4):738–765.e752. doi:10.1016/j.theriogenology.2011.11.012.
98. Aebersold R, Mann M. Mass spectrometry-based proteomics. Nature. 2003;422(6928):198–207. doi:10.1038/nature01511.
99. Breker M, Schuldiner M. The emergence of proteome-wide technologies: systematic analysis of proteins comes of age. Nat Rev Mol Cell Biol. 2014;15(7):453–64. doi:10.1038/nrm3821.
100. Ho YP, Reddy PM. Identification of pathogens by mass spectrometry. Clin Chem. 2010;56(4):525–36. doi:10.1373/clinchem.2009.138867.
101. Fenn JB, Mann M, Meng CK, Wong SF, Whitehouse CM. Electrospray ionization for mass spectrometry of large biomolecules. Science. 1989;246(4926):64–71.
102. Karas M, Hillenkamp F. Laser desorption ionization of proteins with molecular masses exceeding 10,000 daltons. Anal Chem. 1988;60(20):2299–301.
103. Tanaka K, Waki H, Ido Y, Akita S, Yoshida Y, Yoshida T, Matsuo T. Protein and polymer analyses up to m/z 100 000 by laser ionization time-of-flight mass spectrometry. Rapid Commun Mass Spectrom. 1988;2:151–3.
104. Wilm M. Principles of electrospray ionization. Mol Cell Proteomics. 2011;10(7):M111.009407. doi:10.1074/mcp.M111.009407.
105. Yates JR, Ruse CI, Nakorchevsky A. Proteomics by mass spectrometry: approaches, advances, and applications. Annu Rev Biomed Eng. 2009;11:49–79. doi:10.1146/annurev-bioeng-061008-124934.
106. Cunsolo V, Muccilli V, Saletti R, Foti S. Mass spectrometry in food proteomics: a tutorial. J Mass Spectrom. 2014;49(9):768–84. doi:10.1002/jms.3374.
107. Chen CH. Review of a current role of mass spectrometry for proteome research. Anal Chim Acta. 2008;624(1):16–36. doi:10.1016/j.aca.2008.06.017.
108. Dawson P. Quadrupole mass analyzers: performance, design and some recent applications. Mass Spectrom Rev. 1986;5:1–37.
109. Shevchenko A, Loboda A, Shevchenko A, Ens W, Standing K. MALDI quadrupole time-of-flight mass spectrometry: A powerful tool for proteomic research. Anal Chem. 2000;72(9):2132-41. doi:10.1021/ac9913659.
110. Schwartz JC, Senko MW, Syka JE. A two-dimensional quadrupole ion trap mass spectrometer. J Am Soc Mass Spectrom. 2002;13(6):659–69. doi:10.1016/S1044-0305(02)00384-7.
111. Cho WC. Proteomics technologies and challenges. Genomics Proteomics Bioinformatics. 2007;5(2):77–85. doi:10.1016/S1672-0229(07)60018-7.

112. Scigelova M, Hornshaw M, Giannakopulos A, Makarov A. Fourier transform mass spectrometry. Mol Cell Proteomics. 2011;10(7):M111.009431. doi:10.1074/mcp.M111.009431.
113. Eng JK, McCormack AL, Yates JR. An approach to correlate tandem mass spectral data of peptides with amino acid sequences in a protein database. J Am Soc Mass Spectrom. 1994;5(11):976–89. doi:10.1016/1044-0305(94)80016-2.
114. Cash P. Proteomics of bacterial pathogens. Expert Opin Drug Discov. 2008;3(5):461–73. doi:10.1517/17460441.3.5.461.
115. Barbuddhe SB, Maier T, Schwarz G, Kostrzewa M, Hof H, Domann E, et al. Rapid identification and typing of *Listeria* species by matrix-assisted laser desorption ionization-time of flight mass spectrometry. Appl Environ Microbiol. 2008;74(17):5402–7. doi:10.1128/AEM.02689-07.
116. Bernardo K, Pakulat N, Macht M, Krut O, Seifert H, Fleer S, et al. Identification and discrimination of *Staphylococcus aureus* strains using matrix-assisted laser desorption/ionization-time of flight mass spectrometry. Proteomics. 2002;2(6):747–53. doi:10.1002/1615-9861(200206)2:6<747::AID-PROT747>3.0.CO;2-V.
117. Holland RD, Wilkes JG, Rafii F, Sutherland JB, Persons CC, Voorhees KJ, Lay JO. Rapid identification of intact whole bacteria based on spectral patterns using matrix-assisted laser desorption/ionization with time-of-flight mass spectrometry. Rapid Commun Mass Spectrom. 1996;10(10):1227–32. doi:10.1002/(SICI)1097-0231(19960731)10:10<1227::AID-RCM659>3.0.CO;2-6.
118. Schulze WX, Usadel B. Quantitation in mass-spectrometry-based proteomics. Annu Rev Plant Biol. 2010;61:491–516. doi:10.1146/annurev-arplant-042809-112132.
119. Andersen JS, Wilkinson CJ, Mayor T, Mortensen P, Nigg EA, Mann M. Proteomic characterization of the human centrosome by protein correlation profiling. Nature. 2003;426(6966):570–4. doi:10.1038/nature02166.
120. Bantscheff M, Schirle M, Sweetman G, Rick J, Kuster B. Quantitative mass spectrometry in proteomics: a critical review. Anal Bioanal Chem. 2007;389(4):1017–31. doi:10.1007/s00216-007-1486-6.
121. Beynon RJ, Doherty MK, Pratt JM, Gaskell SJ. Multiplexed absolute quantification in proteomics using artificial QCAT proteins of concatenated signature peptides. Nat Methods. 2005;2(8):587–9. doi:10.1038/nmeth774.
122. Carranza P, Grunau A, Schneider T, Hartmann I, Lehner A, Stephan R, et al. A gel-free quantitative proteomics approach to investigate temperature adaptation of the food-borne pathogen Cronobacter turicensis 3032. Proteomics. 2010;10(18):3248–61. doi:10.1002/pmic.200900460.
123. Liu X, Gao B, Novik V, Galán JE. Quantitative proteomics of intracellular *Campylobacter jejuni* reveals metabolic reprogramming. PLoS Pathog. 2012;8(3), e1002562. doi:10.1371/journal.ppat.1002562.
124. Washburn MP, Wolters D, Yates JR. Large-scale analysis of the yeast proteome by multidimensional protein identification technology. Nat Biotechnol. 2001;19(3):242–7. doi:10.1038/85686.
125. Gilmore JM, Washburn MP. Advances in shotgun proteomics and the analysis of membrane proteomes. Journal of Proteomics. 2010;73(11):2078–91. doi:10.1016/j.jprot.2010.08.005.
126. Kudva IT, Stanton TB, Lippolis JD. The *Escherichia coli* O157:H7 bovine rumen fluid proteome reflects adaptive bacterial responses. BMC Microbiol. 2014;14:48. doi:10.1186/1471-2180-14-48.
127. Nilsson RE, Ross T, Bowman JP, Britz ML. MudPIT profiling reveals a link between anaerobic metabolism and the alkaline adaptive response of *Listeria* monocytogenes EGD-e. PLoS One. 2013;8(1), e54157. doi:10.1371/journal.pone.0054157.
128. Gavin AC, Maeda K, Kühner S. Recent advances in charting protein-protein interaction: mass spectrometry-based approaches. Curr Opin Biotechnol. 2011;22(1):42–9. doi:10.1016/j.copbio.2010.09.007.

129. Rigaut G, Shevchenko A, Rutz B, Wilm M, Mann M, Séraphin B. A generic protein purification method for protein complex characterization and proteome exploration. Nat Biotechnol. 1999;17(10):1030–2. doi:10.1038/13732.
130. Butland G, Peregrín-Alvarez JM, Li J, Yang W, Yang X, Canadien V, et al. Interaction network containing conserved and essential protein complexes in *Escherichia coli*. Nature. 2005;433(7025):531–7. doi:10.1038/nature03239.
131. Burnaevskiy N, Fox TG, Plymire DA, Ertelt JM, Weigele BA, Selyunin AS, et al. Proteolytic elimination of *N*-myristoyl modifications by the *Shigella* virulence factor IpaJ. Nature. 2013;496(7443):106–9. doi:10.1038/nature12004.
132. Parker CE, Mocanu V, Mocanu M, Dicheva N, Warren MR. Mass spectrometry for post-translational modification. In: Alzate O, editor. Neuroproteomics. Boca Raton, FL: CRC Press; 2010.
133. Larsen MR, Trelle MB, Thingholm TE, Jensen ON. Analysis of posttranslational modifications of proteins by tandem mass spectrometry. Biotechniques. 2006;40(6):790–8.
134. Schirm M, Schoenhofen IC, Logan SM, Waldron KC, Thibault P. Identification of unusual bacterial glycosylation by tandem mass spectrometry analyses of intact proteins. Anal Chem. 2005;77(23):7774–82. doi:10.1021/ac051316y.
135. Macek B, Mijakovic I, Olsen JV, Gnad F, Kumar C, Jensen PR, Mann M. The serine/threonine/tyrosine phosphoproteome of the model bacterium *Bacillus subtilis*. Mol Cell Proteomics. 2007;6(4):697–707. doi:10.1074/mcp.M600464-MCP200.
136. Kamath KS, Vasavada MS, Srivastava S. Proteomic databases and tools to decipher post-translational modifications. J Proteomics. 2011;75(1):127–44. doi:10.1016/j.jprot.2011.09.014.
137. Bensimon A, Heck AJ, Aebersold R. Mass spectrometry-based proteomics and network biology. Annu Rev Biochem. 2012;81:379–405. doi:10.1146/annurev-biochem-072909-100424.
138. Barrera NP, Robinson CV. Advances in the mass spectrometry of membrane proteins: from individual proteins to intact complexes. Annu Rev Biochem. 2011;80:247–71. doi:10.1146/annurev-biochem-062309-093307.
139. Berrade L, Garcia AE, Camarero JA. Protein microarrays: novel developments and applications. Pharm Res. 2011;28(7):1480–99. doi:10.1007/s11095-010-0325-1.
140. Ray S, Mehta G, Srivastava S. Label-free detection techniques for protein microarrays: prospects, merits and challenges. Proteomics. 2010;10(4):731–48. doi:10.1002/pmic.200900458.
141. Angenendt P, Kreutzberger J, Glökler J, Hoheisel JD. Generation of high density protein microarrays by cell-free in situ expression of unpurified PCR products. Mol Cell Proteomics. 2006;5(9):1658–66. doi:10.1074/mcp.T600024-MCP200.
142. MacBeath G. Protein microarrays and proteomics. Nat Genet. 2002;32(Suppl):526–32. doi:10.1038/ng1037.
143. Templin MF, Stoll D, Schwenk JM, Pötz O, Kramer S, Joos TO. Protein microarrays: promising tools for proteomic research. Proteomics. 2003;3(11):2155–66. doi:10.1002/pmic.200300600.
144. Danckert L, Hoppe S, Bier FF, von Nickisch-Rosenegk M. Rapid identification of novel antigens of *Salmonella* Enteritidis by microarray-based immunoscreening. Mikrochim Acta. 2014;181(13-14):1707–14. doi:10.1007/s00604-014-1197-6.
145. Gehring A, He X, Fratamico P, Lee J, Bagi L, Brewster J, et al. A high-throughput, precipitating colorimetric sandwich ELISA microarray for Shiga toxins. Toxins (Basel). 2014;6(6):1855–72. doi:10.3390/toxins6061855.
146. Gonzalez-Gonzalez M, Jara-Acevedo R, Matarraz S, Jara-Acevedo M, Paradinas S, Sayagües JM, et al. Nanotechniques in proteomics: protein microarrays and novel detection platforms. Eur J Pharm Sci. 2012;45(4):499–506. doi:10.1016/j.ejps.2011.07.009.
147. Zhao Y, Jensen ON. Modification-specific proteomics: strategies for characterization of post-translational modifications using enrichment techniques. Proteomics. 2009;9(20):4632–41. doi:10.1002/pmic.200900398.

148. Rathert P, Dhayalan A, Murakami M, Zhang X, Tamas R, Jurkowska R, et al. Protein lysine methyltransferase G9a acts on non-histone targets. Nat Chem Biol. 2008;4(6):344–6. doi:10.1038/nchembio.88.
149. Schweitzer B, Predki P, Snyder M. Microarrays to characterize protein interactions on a whole-proteome scale. Proteomics. 2003;3(11):2190–9. doi:10.1002/pmic.200300610.
150. Qian X, Metallo SJ, Choi IS, Wu H, Liang MN, Whitesides GM. Arrays of self-assembled monolayers for studying inhibition of bacterial adhesion. Anal Chem. 2002;74(8):1805–10.
151. Chen CS, Sullivan S, Anderson T, Tan AC, Alex PJ, Brant SR, et al. Identification of novel serological biomarkers for inflammatory bowel disease using *Escherichia coli* proteome chip. Mol Cell Proteomics. 2009;8(8):1765–76. doi:10.1074/mcp.M800593-MCP200.
152. Liu CX, Wu FL, Jiang HW, He X, Guo SJ, Tao SC. Global identification of CobB interactors by an *Escherichia coli* proteome microarray. Acta Biochim Biophys Sin Shanghai. 2014;46(7):548–55. doi:10.1093/abbs/gmu038.

Index

X. Deng et al. (eds.), *Applied Genomics of Foodborne Pathogens*, Food Microbiology and Food Safety, DOI 10.1007/978-3-319-43751-4

CPSIA information can be obtained
at www.ICGtesting.com
Printed in the USA
BVOW07*0918100217
475858BV00001B/1/P

9 783319 437491